‖‖‖‖‖‖‖‖‖‖‖‖‖‖‖‖‖‖‖
D1246791

The Mathematics of Sonya Kovalevskaya

ИМПЕРАТОРСКАЯ АКАДЕМІЯ НАУКЪ

НА ОСНОВАНІИ ВЫСОЧАЙШЕ ДАРОВАННАГО ЕЙ ВЪ 8-Й ДЕНЬ ЯНВАРЯ 1836 ГОДА

ИЗБРАЛА

СОФЬЮ ВАСИЛЬЕВНУ КОВАЛЕВСКУЮ

СОСТОЯЩУЮ ПРОФЕСОРОМЪ СТОКГОЛЬМСКАГО УНИВЕРСИТЕТА

ВЪ СВОИ ЧЛЕНЫ-КОРЕСПОНДЕНТЫ ПО РАЗРЯДУ МАТЕМАТИЧЕСКИХЪ НАУКЪ

29 ДЕКАБРЯ 1889 ГОДА

ИМПЕРАТОРСКОЙ АКАДЕМІИ НАУКЪ ПРЕЗИДЕНТЪ

ВИЦЕ-ПРЕЗИДЕНТЪ

НЕПРЕМѢННЫЙ СЕКРЕТАРЬ

Kovalevskaya's certificate of election as a corresponding member of the Imperial Academy of Sciences, dated 29 December 1889 (10 January 1890). The certificate reads: The Imperial Academy of Sciences, by the authority of its supreme mandate of 8 January 1836, has elected Sophia Vasilievna Kovalevskaya, professor of the University of Stockholm, a corresponding member in the section of mathematical sciences. (Courtesy of the Institut Mittag-Leffler.)

Roger Cooke

The Mathematics
of Sonya Kovalevskaya

With 21 Figures

Springer-Verlag
New York   Berlin   Heidelberg   Tokyo

Roger Cooke
Department of Mathematics
University of Vermont
Burlington, Vermont 05405
U.S.A.

Library of Congress Cataloging in Publication Data
Cooke, Roger,
    The mathematics of Sonya Kovalevskaya.
    Bibliography: p.
    Includes index.
    1. Mathematical analysis—History.   2. Kovalevskaia, S. V. (Sof'ia
Vasil'evna), 1850–1891.
    3. Mathematicians—Soviet Union—Biography.
I. Title.
QA300.C65   1984   515'.09   84-10599

Typeset by Interactive Composition Corporation, Pleasant Hill, California.
Printed and bound by R. R. Donnelley and Sons, Harrisonburg, Virginia.
Printed in the United States of America.

9 8 7 6 5 4 3 2 1

ISBN 0-387-96030-9 Springer-Verlag New York Berlin Heidelberg Tokyo
ISBN 3-540-96030-9 Springer-Verlag Berlin Heidelberg New York Tokyo

To Cathie, who keeps me in touch with what is most important,
To Katya, Andrew, and Hilary, who bring us joy every day,
To my father, who gave me an education beyond books and classrooms,
                    and
In loving memory of my mother

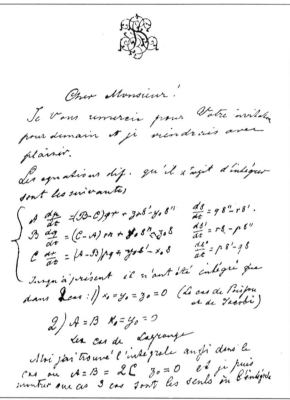

Letter from Kovalevskaya communicating her discovery of the Kovalevskaya case of a rotating rigid body. The letter was in Mittag-Leffler's possession and is possibly to him, though she usually addressed him less formally at the time the letter was written. (Courtesy of the Institut Mittag-Leffler.)

Preface

This book is the result of a decision taken in 1980 to begin studying the history of mathematics in the nineteenth century. I hoped by doing it to learn something of value about Kovalevskaya herself and about the mathematical world she inhabited. Having been trained as a mathematician, I also hoped to learn something about the proper approach to the history of the subject. The decision to begin the study with Kovalevskaya, apart from the intrinsic interest of Kovalevskaya herself, was primarily based upon the fact that the writing on her in English had been done by people who were interested in sociological and psychological aspects of her life. None of these writings discussed her mathematical work in much detail. This omission seemed to me a serious one in biographical studies of a woman whose primary significance was her mathematical work. In regard to both the content of nineteenth century mathematics and the nature of the history of mathematics I learned a great deal from writing this book. The attempt to put Kovalevskaya's work in historical context involved reading dozens of significant papers by great mathematicians. In many cases, I fear, the purport of these papers is better known to many of my readers than to me. If I persevered despite misgivings, my excuse is that this book is, after all, primarily about Kovalevskaya. If specialists in Euler, Cauchy, etc., find omissions or misinterpretations in my handling of the works of these authors, I can only plead that the background of a painting never has the same clarity as the foreground. In reading secondary sources I found noticeable differences between the accounts written by mathematicians and those written by historians. Among other differences, the former devote much more space to the technical details of how a result was obtained, while the latter tend to emphasize the influence of one idea or school upon another. As a newcomer to the field, I considered it my place to learn as much as I could from both mathematicians and historians. Naturally, the present work is written in the "mathematician's" style, though I have tried to make it as accessible as possible to nonspecialists.

Since Kovalevskaya's periods of mathematical activity, first as a student of Weierstrass and later as a mature scholar, were separated by a 6-year period of more or less domestic life, this natural division has been used as the plan for this book. In order to fix the chronology, the chapters discussing her work during these two periods are preceded by chapters of biography. It is very light biography, written to entertain as well as to instruct. It skims lightly over the years, touching down at points that seemed particularly interesting. The reader should by no means infer that the incidents reported here for any given year were necessarily the most important ones for understanding Kovalevskaya's life. For a serious scientific biography of Kovalevskaya the reader should turn to the excellent recent account by Ann Hibner Koblitz

(1983). Readers who know Russian will also find much of interest in the book of P. Ya. Kochina (1981). The present book uses the two just mentioned extensively as sources. The primary sources used in Chapters 1, 5, and 9 of the present book are in the archives of the Institut Mittag-Leffler. All letters quoted in this book, unless otherwise cited, come from that source.

The motivation for this book resides in the six chapters of mathematical analysis, in which an attempt is made to place Kovalevskaya's work in context within the history of nineteenth-century analysis. The effect of focusing on, and, so to speak, enlarging Kovalevskaya's papers, is a certain distortion of the picture. To make this book of reasonable length and readability, it was necessary to restrict the discussion of the works of her predecessors to the most significant results relevant to Kovalevskaya's work. Such a restriction makes the history of each topic resemble a ladder on which each successive mathematician moved one rung higher than his predecessors, whereas the reality is more like a tree, with branches sprouting in all directions. With that caveat I trust the reader will be better able to place Kovalevskaya's work in perspective after reading this book than before.

A few words are in order about the text itself, especially the mathematical analysis in Chapters 2–4 and 6–8. These chapters contain too few formulas to constitute a textbook exposition of the subjects they discuss and too many for a popular exposition. My impression after reading older histories of mathematics which discuss the works of great mathematicians in prose has been that such a style conveys no real idea of what the mathematics was. The formulas seem essential to any real communication of ideas. They should be used however as one would use a child's model airplane to explain that invention to someone who had never seen one. One would not expect to fly in a model airplane. For the same reason, one should not take the formulas in this book, even the long strings of formulas, for an attempt at a mathematical proof. The formulas are in the book only as a means of referring to the mathematics.

The six appendices were written for two reasons. The first was to make the book more "user-friendly" to a hypothetical undergraduate reader who has had only a year or two of mathematics. With these appendices, such a reader should be able to understand the gist of what is said in the text. The second reason for the appendices is that in the text it was necessary to touch on two topics (Weierstrass' method of solving partial differential equations of a certain form, and Jacobi's last multiplier method) which are no longer part of the basic courses in differential equations. I thought that mathematicians who, like me, had not seen these topics before, would be interested in a more detailed exposition of them.

Finally I should also explain my system, or rather my lack of system, in transliterating Cyrillic. Since the ending -sky is familiar to Europeans and Americans in such names as Dostoevsky and Tchaikovsky, I see no point in using -skii. In general Russian words are transliterated in a way which will cause the average English-speaking reader to pronounce them in an approximately correct manner. Those who know Russian will recognize these words anyway, and those who do not will suffer no harm from reading a nonstandard transliteration. Knowing that Kovalevskaya's maternal ancestors came from Germany, I refer to them by the name Schubert, rather than Shubert. On the other hand, I have retained the spelling Mendel'son in preference to Mendelssohn, because the woman in question was Polish. In connection with Russian names, many conversations with non-Russian-speaking devotees of Russian literature have convinced me that the invariable sequence of given-name-plus-patronymic is excessively wearisome to most readers. Therefore I make very little use of patronymics, trusting that any Russian-speaking readers will understand that no disrespect is intended. Similarly, in order to avoid confusing my readers with the many diminutives of the name Sophia I consistently use "Sonya" even though Kovalevskaya did not get that nickname until adulthood.

It is a pleasure to record my gratitude to the many organizations and individuals who have assisted me in carrying out this project. My sincere thanks go

to Patricia and Philip Kitcher for many years of friendship and for advice on both technical and scholarly problems;

to the University of Vermont for sabbatical leave and travel support;

to Lennart Carleson, Mrs. Karin Göransson, and the Institut Mittag-Leffler for allowing me to use the archives of the Institut and for arranging housing in Djursholm, Sweden for my family and me in the summer of 1981, a summer we shall always remember as one of the most pleasant of our lives;

to the International Research & Exchanges Board for nominating me for the 1981–82 exchange with the Soviet Union and for emergency support when the vagaries of international cooperation made my trip to the Soviet Union impossible;

to Ivor Grattan-Guinness for invaluable help; I trust he will accept this acknowledgment in lieu of a citation at each point where I have used one of the references he supplied; he will know from comparing this book with the draft he read how many improvements are due to his advice;

to the London Mathematical Society, the British Society for the History of Mathematics, University College London, and the British Museum for access to sources and other assistance during my year in England. Indeed I would like to thank the entire United Kingdom for making the year 1981–82 so pleasant for me and my family;

to Jeremy Gray for helpful conversations;

to P. Ya. Kochina for helpful correspondence;

to Ann Hibner Koblitz for sharing information, thereby saving me many mistakes;

to Clark University, Worcester, Massachusetts for permission to consult archival materials;

to the staff of Springer-Verlag for their very considerate cooperation.

Roger Cooke

Contents

PART I

CHILDHOOD AND EDUCATION

CHAPTER 1
Biography:1850–1874

1.1. Introduction

Sophia Vasilievna Kovalevskaya (1850–1891), whose mathematical works will be analyzed in the chapters which follow, is becoming increasingly well known as an early woman of achievement. As the feminist movement inspired a new look at history and an attempt to appreciate the formerly neglected contributions of women to all aspects of culture, Kovalevskaya was inevitably recognized as one of the first women to make contributions of high quality to mathematics, which even today is predominantly a "male" subject. The result of this increased interest is a number of recent articles and books on Kovalevskaya which tell of various aspects of her life and character. At the same time certain questions of psychology and sociology have arisen on which no consensus has yet been reached. For example, are there innate differences between men and women which cause women to gravitate toward literary activity and men toward quantitative activity? Will women constitute a sizeable portion of our scientists in the near future? Kovalevskaya has become part of the debate, as various claims have been made for her and attempts have been made to evaluate her many activities. Although this book is primarily an attempt to understand Kovalevskaya's contributions to mathematics, it may perhaps contribute to the discussion by recounting her work accurately in the context of the history of mathematics. The present chapter, which sketches Kovalevskaya's biography to the year 1874, and Chapter 5, which tells the story of the rest of her life, are intended to provide a chronological framework for the mathematical Chapters 2–4 and 6–8.

Conveniently for the biographer, Kovalevskaya's mathematical activity divides nearly into two periods, the first being her work as a student of Weierstrass and the second her work as a mature scholar. The two parts are separated by a turbulent 6 years of domestic life, mostly in Russia. This division provides the plan of the book. The present chapter discusses her life up to the time when she wrote her doctoral dissertations and will be followed by three chapters of analysis of these dissertations. These four chapters make up Part I. Part II begins with Chapter 5, which tells the story of the last 16 years of Kovalevskaya's life and is followed by four chapters analyzing her mathematical work. Part III consists of six appendices which discuss technical mathematical points of the text, followed by the bibliography.

The best detailed biographies of Kovalevskaya are those of Ann Hibner Koblitz (1983) and P. Ya. Kochina (1981). Kovalevskaya's autobiography of the first 15 years of her life (1890d) together with her mathematical autobiography (1890c) were translated into English and published by B. Stillman

(1978). An earlier translation of the former is Leffler (1895). In addition many short articles on Kovalevskaya exist, for example, Tee (1977), Rappaport (1981), Kramer (1973), Mittag-Leffler (1923c), and Bell (1937). A source book on Kovalevskaya was compiled by Shtraikh (1951), and a compilation of her literary works was made by Kochina (1974). In addition Kochina has published Weierstrass' letters to Kovalevskaya both in German and in Russian translation (1973) and Russian translations of some of her correspondence with other people (1979), as well as the Russian edition of her collected works (*Raboty*). More references can of course be found in these works.

1.2. Kovalevskaya's Ancestors

Kovalevskaya was born Sophia Vasilievna Krukovskaya in Moscow on 15 January 1850 (3 January on the calendar then in use in Russia). According to Kochina (1981, pp. 7–8) the records of the Church of the Apparition beyond the Petrine Gates for 1850 contain the following entry:

> Born 3 January, baptized 17 Sophia; her parents are Colonel of the Artillery Vasily Vasilievich Krukovskoi and his legal wife Elizaveta Fyodorovna; the husband is of the Orthodox faith, the wife Lutheran. Godparents are retired Second Lieutenant of the Artillery Semyon Vasilievich Krukovskoi and Anna Vasilievna Krukovskaya, daughter of Proviantmeister Vasily Semyonovich Krukovskoi. The sacrament of baptism was administered by parish priest Pavel Krylov with deacon Pavel Popov and sacristan Alexander Speransky.

It appears from this entry that Kovalevskaya's godparents were her father's brother and sister, all three being children of Vasily Semyonovich Krukovskoi. The baptismal record caught the family name about midstage in a curious metamorphosis which it underwent during the 1840s and 1850s. According to Koblitz (1983, p. 9), Kovalevskaya's paternal grandfather was a Russianized Pole who owned an estate in the Pskov region. Since the baptismal record refers to him as proviantmeister, we may infer that he had served in the army. He wrote the family name as Kriukovskoi with the stress on the last syllable. According to Kochina (1981, pp. 7–8; 1974, pp. 509–510), during the 1850s the family applied for recognition as members of the nobility, perhaps in order to secure certain privileges in case the serfs were emancipated. In 1858 the family was granted such recognition and listed in section six of the genealogical register as descended from the family Krukovsky (stress on the middle syllable). In addition the claim of descent from a Hungarian king was recognized, and the family was allowed to call itself Korvin-Krukovsky.

Except for these strenuous efforts to link themselves with the nobility the Kriukovskoi family do not seem to have done anything memorable. Kovalevskaya's maternal ancestry, however, is much more interesting. The following account is given by Kochina (1981, pp. 10–11). Kovalevskaya's great-great-grandfather, Johann Ernst Schubert, was a Lutheran theologian who was forced to move from Braunschweig to Pomerania as a result of the Seven Years' War. The youngest of his six sons, Theodor (Kovalevskaya's great-grandfather) studied theology and eastern languages at the University of Göttingen. For unknown reasons he emigrated to Russia, becoming a surveyor in Revel (now Tallinn in Estonia). In Russia his name became Fyodor Ivanovich ("Theodor, son of Johann") Schubert. Through independent study he became well versed in mathematics and astronomy and within a few years became a member of the Petersburg Academy of Sciences, in charge of the Kunstkammer of the Academy together with the observatory and the library. He corresponded with Gauss, Laplace, Bessel, and other great mathematicians.

Kovalevskaya's grandfather, Fyodor Fyodorovich Schubert had a distinguished scientific and military career, publishing many works on geodesy and geographical maps of Russia. In 1827 he became an honorary member of the Petersburg Academy of Sciences, and his 1858 book on astronomical and geodesic work in Russia, published in French, received high acclaim in Western Europe. He lived on Vasilievsky Island in Petersburg, and there he raised his son (another Fyodor Fyodorovich Schubert) and three daughters, Elizaveta (Kovalevskaya's mother), Alexandrina, and Sophia.

Thus, as far as the limited data available to us can be studied, it appears that Kovalevskaya's paternal ancestors were respectable but otherwise undistinguished lower gentry living in the western part of European Russia, while her maternal ancestors were descended from German immigrants with a tradition of vigorous intellectual activity. We are of course confined by lack of data to her male ancestors.

1.3. Kovalevskaya's Family

Kovalevskaya's parents, Vasily Vasilievich Kriukovskoi (1800–1874) and Elizaveta Fyodorovna Schubert (1820–1879), were married on 29 January 1843 (17 January on the Russian calendar). Judging from the accounts of Koblitz and Kochina their married life must have resembled somewhat the lives of the aristocratic couples found in the novels of Tolstoy and Dostoevsky. Elizaveta had been raised in society and was happiest when attending a ball or an intimate salon where she could talk about the currently fashionable topic of conservation. She was an accomplished pianist and amateur actress. Vasily was a military man, who liked nothing better than playing cards at his club. In true Russian-novel style he lost huge sums of money and had to sell part of his estates in order to cover his gambling debts. How this conflict of interests would be resolved in a society where wives had no rights to speak of can easily be imagined. Elizaveta's diaries from the periods 1843–1851 and 1863–1864 have been preserved, and the former are especially full of bitter complaints against her uncaring, indifferent husband (Kochina 1981, pp. 12–13). Vasily was 20 years older than his wife and by

all accounts treated her as a child. Such an attitude on his part could not have
fostered respect for their mother among the children, nor did Elizaveta attempt to brighten her drab existence by taking an interest in the children. (Her
principal amusement seems to have been a home theater in which she played
leading roles.) Her diaries from 1843 to 1851 make no mention of either of
the two daughters she bore during this time. Kovalevskaya's reminiscences
of her childhood hardly mention her mother, but they do confirm the impression that her mother's role in the family was not considered significant
(Stillman 1978, p. 109). Kovalevskaya said that she always felt clumsy on the
rare occasions when she tried to show affection to her mother. As Koblitz
remarks (1983, p. 29) her mother's remoteness and general unimportance to
the family had one very important consequence for Kovalevskaya's development. They weakened her mother's influence during her adolescence, the
time when most girls her age were learning social graces and putting aside all
interests except planning for marriage.

Kovalevskaya's father was a well-educated man who spoke English and
French fluently. He pursued a conventional military career through the last
years of the reign of Alexander I and all through the reign of Nikolai I,
starting as a Junker in 1817 and by 1848 rising to the rank of colonel of the
artillery and commander of the Moscow artillery garrison. While his future
father-in-law Fyodor Fyodorovich Schubert may have been involved in the
Decembrist revolt which followed the death of Alexander I in 1825, Vasily
definitely avoided such entanglements. He went on the Balkan campaign
during the Turkish war of 1828–1830 and was decorated several times. By the
time of his retirement in 1858 he had attained the rank of lieutenant-general.
Upon his retirement in 1858 the family went to live on his estate at Palibino
(now Polibino, about midway between Leningrad and Kiev). It is believed
(Kochina 1981, p. 15) that Vasily felt the need for prudently overseeing his
estates, as rumors were in the air that the serfs were about to be emancipated.
The accession of Alexander II in 1855, soon followed by Russia's humiliation
in the Crimean War, had led to a general disillusionment with conservatism
in Russia and an expectation of reform. Vasily may have been attracted by
the change in career. In 1863 he was elected Marshal of the Nobility for the
province of Vitebsk, an office which had been created at the time of the
constitutional reform under Catherine the Great. With the emancipation of the
serfs in 1861 a new system of provincial administration was needed. This
need was met by organizing local and provincial assemblies called zemstva
(the word is plural, singular zemstvo) with representation from the peasants,
the townspeople, and the nobility, heavily biased in favor of the last of these.
The president of the provincial zemstvo was by law the Marshal of the
Nobility. Thus Vasily occupied a rather important position at a time and place
of great historical importance: near the Polish border at the time of the Polish
uprising of 1862–1863. By steering a cautiously neutral course he was able
to retain his office when all those who held the same office in neighboring
provinces had been sacked. According to Kovalevskaya (Kochina 1981, p.

31) her father did not believe the rebellion had a chance of success or that it would be good for the people even if it did. (In a sense he was right. Emancipation of the serfs in Poland under the Russian regime was carried out on terms much more favorable to the peasants than had been done in Russia and, it seems likely, more favorable than any Polish government would have granted.) Nevertheless his Polish ancestry and his apparent refusal to take an active part in suppressing the rebellion made his position rather precarious for a while.

About a year after her marriage, Elizaveta gave birth to a daughter, Anna Vasilievna, Kovalevskaya's older sister, who, as she developed, fitted very well the current ideal for girls. She was blonde and healthy, naturally attractive, and quite at ease in society. It is natural to suppose that, having a near-perfect daughter, Vasily hoped his next child would be a boy. According to Kochina (1981, p. 12), Elizaveta did bear a son, Vasily Vasilievich, who died young. Kochina does not give the date of his birth, but he almost certainly preceded the surviving son (born in 1855) since he was named after his father. Whatever the case, the next surviving child was Sophia, born in 1850 and quite a contrast to her fair sister. She had a dark complexion and a very intense and serious personality. As mentioned previously, her relations with her mother seem always to have been rather awkward, whereas Aniuta, as Anna was called, related naturally and affectionately to her mother. "Sonya" always felt closer to her father, perhaps because of the serious temperament she shared with him.

Part of Sonya's first eight years of life were spent in Kaluga, south of Moscow, where her father was stationed with the artillery. The children were in the care of a nanny, of course, and Elizaveta was trying to find some outlet for her talents by giving concerts on the piano. Kochina (1981, p. 14) reports one concert of 8 September 1855 for which Elizaveta received high praise. In the nursery Sonya was the nanny's favorite; and her conviction that she was not loved, not having the charm of her sister, and not being the son and heir her father wanted, was reinforced by Nanny's calling attention to every slight which she felt had been inflicted on little Sonya.

In the pivotal year of 1858 the family received the right to call itself Korvin-Krukovsky and was entered in the genealogical register. That same year Vasily Krukovsky retired and took his family to live at Palibino. There, for the first time, free of his military obligations and taking an interest in the management of his household, he discovered that his daughters were appallingly ignorant. He replaced the nanny with an English governess, Miss Margaret Smith, and a Polish tutor, Joseph Malevich. Aniuta, who was already 14, was quite incorrigible; consequently Sonya was the principal beneficiary of the new regimen. Her natural disinclination to socialize had turned her into a moody and introspective 8-year-old with a passionate love of reading. Miss Smith forbade all books except those specifically approved by her and insisted that Sonya take a daily "constitutional" walk with her. (This practice suffered a sudden check one day when the walkers chanced upon a mother bear with two cubs.) Sonya had to read books on the sly when Miss Smith believed she was taking indoor exercise in the library.

1.4. Kovalevskaya's Uncles

Among the members of the extended family who formed the environment of Kovalevskaya's childhood, two uncles received special mention in Kovalevskaya's autobiography. One of these was her mother's younger brother Fyodor Fyodorovich Schubert, who occasionally took Sonya on his knee and told her stories, not the usual Russian fairy tales about firebirds and helpful wolves, but real biology, about infusoria and algae. It may be doubted whether these lessons in limnology were of great importance to the child, but the opportunity to have the undivided attention of her idol was important to her.

On her father's side also Sonya had an uncle who awakened her interest in scientific things. This uncle, Pyotr Vasilievich Krukovsky, had served in the army long enough to become a lieutenant before retiring. His wife had been a cruel mistress of the household who was murdered by the servants. Pyotr turned all his property over to his son and reserved for himself only a small pension. Though not highly educated he had a deep respect for mathematics. Kovalevskaya reported that long before she was old enough to know what the words meant he told her about squaring a circle, about asymptotes, and about many other mathematical concepts.

1.5. Aniuta

From Koblitz' account it is clear that Kovalevskaya's real idol during her childhood was her sister Aniuta. Merely because she was 6 years older Aniuta was naturally competent to do things which were beyond Sonya, and of course her natural charm must have aroused Sonya's admiration.

Under Miss Smith's strict rules Sonya was not to be contaminated by contact with her sister. Even so, Miss Smith began watching Aniuta with extraordinary zeal. It seems Aniuta had abandoned the frivolous preoccupation with fine cothes and balls that was expected of girls her age. She had turned into a serious young woman who spent her money on books and her time giving lessons to the peasant children. Aniuta's new seriousness was the result of contact with two young men who had been exposed to the intellectual ferment of the time (Koblitz 1983, p. 35). One of these, Mikhail Ivanovich Semevsky, was a young officer and former pupil of Malevich, who met Aniuta during a visit to his former tutor. He and Aniuta had agreed to be married, but Vasily Krukovsky intervened and forbade Mikhail to visit Palibino. Aniuta soon found solace in the local priest's son, who had studied natural science in Petersburg and returned to Palibino full of the ideas of Darwin. He was socially beneath Aniuta, and her father treated him with condign disdain. From then on the conflict between Aniuta and her father grew ever more acrimonious. Miss Smith was firmly on Vasily's side, but eventually she resigned in frustration. Elizaveta, hoping for peace in the family, accepted the resignation.

Peace did not come with Miss Smith's departure, however; for Aniuta's struggle for independence had only begun. She and her father had begun to argue the merits of Dostoevsky, whose short-lived journal "The Epoch" came to the Krukovsky home. What Vasily Krukovsky did not know at first was that Aniuta had sent Dostoevsky two of her own short stories. Although Aniuta had taken the precaution of conducting her correspondence with Dostoevsky through the housekeeper (who usually managed to keep suspicious letters away from Aniuta's father), by a perverse misfortune Dostoevsky's letter with payment of some 300 rubles for Aniuta's stories arrived on Elizaveta's nameday (5 September 1864) when the housekeeper was too busy to get to the mailbag ahead of Vasily. Vasily noticed the registered letter addressed to the housekeeper and insisted that she open it in his presence. The old general's first reaction was fully in accord with his conservative and patriarchal instincts. Aniuta, however, was not to be intimidated; and the general, a reasonable man at bottom, eventually changed his attitude after hearing the stories read. Through her firmness Aniuta won permission for Dostoevsky to visit her during the winter season in Petersburg. It is likely that Dostoevsky's interest in Aniuta was more than literary, as evidenced by the fact that he paid her for her stories at a time when he could ill afford to do so. He was rather lonely at the time, having lost his brother and his first wife. While he was courting Aniuta, he also won Sonya's heart. She recalled how hard she worked to learn his favorite piano work, Beethoven's Pathetique Sonata, only to find that he ignored her performance of it because he was whispering his love to Aniuta. There is considerable evidence that Dostoevsky proposed marriage to Aniuta but was rejected. Aniuta felt, no doubt correctly, that she would never be free to act independently if she married Dostoevsky.

1.6.   Early Mathematical Training

Perhaps because Aniuta had preempted the literary field, and perhaps because of a natural bent toward science which was reflected in her attachment to her two scientifically minded uncles, Sonya turned her bright mind toward mathematics and biology. In her autobiography she tells how she used to stare fascinated at the walls of one room in the Palibino estate, which for lack of proper wallpaper had been covered with the lithographed notes of a calculus course. Kochina believes this course must have been one given by the famous Russian mathematician Ostrogradsky. Kovalevskaya apparently believed that this haphazard introduction to the subject nevertheless gave her a familiarity with the symbols and formulas of calculus.

A better indication of her talent for mathematics came when she read a book by a friend of the family, the physicist Nikolai Nikanorovich Tyrtov.

This book used trigonometry, a subject unfamiliar to Sonya and also to her tutor. As a result she was forced to make sense of the sine function all by herself. Her guess was that the sine of a central angle in a circle is proportional to the chord subtended by the angle. In fact it is proportional to half the subtended chord of twice the angle, but for small angles the difference is negligible. Indeed the earliest trigonometric tables–in Ptolemy's *Almagest,* for example–were tables of chords rather than half-chords. Thus Kovalevskaya was following the route of the pioneers in the area. Tyrtov, when he realized what Sonya had done, called her a "new Pascal" and urged her father to allow her to study higher mathematics. Vasily Krukovsky finally consented to allow Sonya to gain at least a dilettante's appreciation of mathematics by taking lessons from Professor Alexander Nikolaevich Strannoliubsky in Petersburg.

There are some puzzling aspects to Vasily Krukovsky's decision to allow his daughter to study mathematics. In her autobiographical essay on her mathematical career (Stillman 1978, pp. 213–229) Kovalevskaya wrote that he harbored a strong prejudice against learned women. Besides, Strannoliubsky was a known liberal who had served as inspector for a tuition-free school on Vasilievsky Island until the school was closed in 1866. It is therefore curious that Tyrtov should recommend Strannoliubsky to a conservative like Vasily Krukovsky. Perhaps as Koblitz suggests (1983, p. 47), the best explanation is Sonya's own fierce determination.

1.7.   In Search of Higher Education

With Strannoliubsky, Sonya studied differential and integral calculus. Time was moving forward, however, and for Aniuta, who was now 23, and Sonya, who was 17, the time had come to try to put the foundation under the castles they had built in the air. There was no hope of higher education or a career in Russia since the liberalizing movement of 1861–1862 had come to an end. Consequently, like many young women of their generation, they decided to go to the west. The problem, a formidable one for a woman in Russia, was how to get there. Women were not ordinarily issued passports in their own names. A dependent woman had to have permission from her husband or father in order to travel abroad; and women who traveled alone, especially young ones, were bound to be treated with curiosity and suspicion. Like Vasily Krukovsky, most fathers of the time were unwilling to give their daughters permission to study abroad. The solution resorted to in extreme cases by young women was to get their names on the passport of a liberal "husband" who would allow them to pursue their own interests. The word "husband" is in quotation marks because, of course, a real marriage was not intended by either party. The social duties as hostess or mother normally expected of a wife would have been a severe impediment to the pursuit of the interests for which the woman was contracting the marriage in the first place; and men who wanted a real marriage preferred to choose the wife themselves.

One can see what the "wife" expected to gain from the fictitious marriage; what was in it for the "husband" is not clear. Both parties to such a marriage were committing a serious sin of sacrilege in the view of the Orthodox Church, but it is not recorded that any of them was troubled by scruples of that sort. Though the fictitious marriage seems a drastic solution to the problem, one must remember the explanation attributed to E. F. Litvinova, "You don't have to be a genius to understand that, but you do have to be a Russian."

As can well be imagined, the supply of prospective wives for fictitious marriages exceeded the supply of prospective husbands, but the solution to this secondary problem was simple. One of a group of young women would get "married" and then the "couple" would take the entire group to the west, acting as chaperones. Accordingly Aniuta, Sonya, Aniuta's friend Anna Mikhailovna ("Zhanna") Evreinova, and Zhanna's cousin Julia Lermontova (whose father was second cousin to the famous poet Lermontov) proposed marriage to a young professor in Petersburg in the winter of 1867–1868 but were turned down. Not at all discouraged, they simply proceeded to the next name on their list of prospects. Eventually they came to the name of a young publisher, Vladimir Onufrievich Kovalevsky.

1.8.   Vladimir Kovalevsky

The man who eventually became Sonya's husband was born 14 August 1842 to a family of lesser gentry in Vitebsk province, not far from Palibino. He received a thorough grounding in foreign languages and in his last years at school earned money by translating books (Kochina 1981, p. 37). Besides his very quick mind, Vladimir was distinguished for his leaning toward liberal causes. In 1861 he started to work for the department of heraldry in the Senate, but soon asked for permission to travel abroad for his health. He made the rounds of Heidelberg, Tübingen, Paris, and Nice, visiting old classmates and finally arriving in London, where he gave lessons to the daughter of the exiled radical Herzen. By meeting Herzen he guaranteed himself steady surveillance by agents of the tsar.

When he returned to Russia and his bookselling business, he published many scientific texts and, in 1866, Herzen's "Who Is to Blame?" which brought him serious financial loss when the entire printing was burned at the order of the censors.

Anyone who knew Vladimir's past would have considered him an unlikely prospect as a fictitious husband. He had been engaged to a young radical woman, Masha Mikhailis (cf. Koblitz 1983, p. 70), but the engagement had been broken at the last moment. This broken engagement cast suspicion on him among the radical crowd when Herzen's report of Kovalevsky's arrest in April 1866 turned out to be erroneous. A few days after the breaking of this engagement he received a proposal of fictitious marriage from another woman, but he refused. Nevertheless it turned out that

Vladimir was not at all averse to becoming a fictitious husband. After he met Sonya, he expressed a preference for "marrying" her. As in the case with Dostoevsky and Aniuta, it is likely that Vladimir was slightly infatuated with Sonya and unable to keep his mind fixed on what the women regarded as the central issue.

There was considerable resistance to the marriage on the part of Sonya's father, which apparently Sonya overcame by sending word to a distinguished company of guests at Palibino that she and her fiancé were busy with plans for their wedding. Faced with the necessity either to admit to his guests that his daughter was in rebellion or to give public approval to the engagement, Vasily Krukovsky capitulated. Although he had announced the engagement, however, he was reluctant to go through with the marriage. According to Kochina (1981, p. 44), Sonya simply went to Vladimir's apartment and told her mother she would not leave until her parents agreed to allow the marriage to take place.

Sonya and Vladimir were married in September 1868; henceforth Sonya would be known as Frau von Kowalewsky when writing in German and as Mme. de Kowalevsky when writing in French. In this book she will be referred to as Sonya when she is discussed together with Vladimir, otherwise as Kovalevskaya, which is the Russian feminine form of Kovalevsky. She was rather embarrassed by her new name and blushed whenever anyone called her *Gospozha Kovalevskaya,* probably because she knew that others were assuming her marriage was a real one. Koblitz (1983, p. 74) has commented that the fictitious marriage must have seemed a drastic step to Kovalevskaya. Her father had never been completely intransigent, and she must have suspected that in time he would have relented and allowed her to study abroad. The feeling that she had unnecessarily placed herself under obligation to a man she barely knew was likely to make her difficult to live with at times when it seemed that the goal for which she had taken this step might not after all be reached.

1.9. Study Abroad

The plan for the newlyweds to chaperone Aniuta, Zhanna, and Julia to the West went awry, as Zhanna's parents refused to allow her to go and Aniuta's permission was also delayed. The couple lived in Petersburg for a few months, attending lectures on a wide variety of subjects. Sonya finally realized that she would have to restrict her studies to a manageable amount and began to concentrate on mathematics.

Early in 1869 Sonya, Vladimir, and Aniuta left Russia and traveled to Vienna. Aniuta, who wanted most of all to join the politically active young people of the time, traveled on alone to Paris, having arranged to conceal her whereabouts from her parents by sending her mail to Sonya for dispatch to Russia. Vladimir and Sonya moved on to Heidelberg, where Vladimir had visited 3 years earlier. Sonya had written at great length to Aniuta, even before her marriage, about her dream of an ascetic study in Heidelberg. After strenuous efforts by both of them, the administration agreed that Sonya could attend courses with the permission of the professors. Sonya, and later her friend Julia Lermontova, studied with professors whose names are familiar to every high school physics student today: Bunsen, Kirchhoff, and Helmholtz. Sonya studied mathematics with Professors P. du Bois-Reymond and Koenigsberger. It was a rather heavy schedule, 22 hours per week, of which 16 was mathematics. Considering that Sonya probably knew very little beyond calculus, this study must have been exhausting. Nevertheless the newlyweds were quite happy and found considerable time to travel.

On one significant journey to London (October 1869), Vladimir went to visit his acquaintances Thomas Huxley and Charles Darwin. [Vladimir had published a Russian translation of Darwin's work "Variation of Plants and Animals under Domestication" in May of 1867 from proofs given to him by Darwin. The English original of this work did not appear until January 1868 (Koblitz 1983, p. 90).] Meanwhile Sonya, a somewhat exotic personage due to her field of study, was invited to attend one of George Eliot's Sunday salons. Many years later when she wrote her reminiscences of George Eliot (Kovalevskaya 1886) she recalled that Eliot had deliberately led her into a debate on woman's capacity for abstract thought with a man who turned out to be Herbert Spencer. Since these reminiscences have apparently never been translated into English, it may be worthwhile to record Kovalevskaya's impression of George Eliot. She reported that she had criticized Eliot for conveniently causing the death of the leading characters in her novels just when the reader wonders if they will be able to sustain the promising beginnings they have made. Eliot replied that in fact this is when people tend to die in real life. Kovalevskaya remarked that this principle seemed to have applied to Eliot herself.

At this point I would like to note what is probably only a coincidence, but nevertheless is somewhat intriguing. Kovalevskaya later became famous for showing that theta functions could be used to express the solutions of the equations of motion of a rotating irregular body. She wrote in a letter of 1881 (quoted in Section 5.4 of this work) that she had been interested in this problem almost from the beginning of her mathematical studies. In view of that remark the following passage from Eliot's *Middlemarch* is curious:

In short, woman was a problem which, since Mr. Brooke's mind felt blank before it, could hardly be less complicated than the revolutions of an irregular solid (*Middlemarch,* chapter IV).

Since Middlemarch was written between 1869 and 1872, it is possible that Eliot obtained this comparison from Kovalevskaya.

As mentioned above, Kovalevskaya studied for 1 year with du Bois-Reymond and Koenigsberger, who were both students of Weierstrass. Koenigsberger is now mostly forgotten, except for his connection with Kovalevskaya and Weierstrass and possibly his biography of Helmholtz

(1906). Du Bois-Reymond is still remembered for work in partial differential equations (cf. Nový 1971), for work in calculus of variations (du Bois-Reymond 1879; cited, e.g., in Bliss 1925), and for having published Weierstrass' example of a continuous function which does not have a derivative at any point (du Bois-Reymond 1875; cited, e.g., in Riesz-Nagy 1955). According to Osgood (1928, p. 30) he was one of the first to recognize and emphasize the importance of the Cauchy convergence criterion (du Bois-Reymond 1882).

Life in Heidelberg was turning out to be less idyllic than Sonya had imagined it in the rapturous letter she had written to Aniuta. For one thing, she and her husband were living together, though not as husband and wife. They were sharing a platonic ménage à trois with Julia Lermontova. Both parties to this marriage were highly strung, but Sonya was single-mindedly devoted to the attainment of a goal, while Vladimir was apparently not strongly committed to anything at this time. Nevertheless he was interested in the study of biology and geology which Sonya had encouraged him to take up, so much so that Sonya resented his ability to be completely happy with a book and a cup of tea. He may have been less committed to Sonya's emancipation than he made appear when he married her. A sardonic poem (undated) by Sonya called "The Husband's Complaint" (Kochina 1974, pp. 319–320) very likely reflects her feelings at this time. In the poem the husband-narrator tells how shocked he was to find that his wife really meant what she had said before marriage about continuing her studies. He—well, who could blame him? he asks—had carelessly assured her that she could do so, and now she was making his life miserable by holding him to his promise.

Sonya began to require an extraordinary amount of attention. She developed phobias and required constant companionship, thus preventing Vladimir from pursuing his studies in the field. The couple was plagued by money problems, since Vladimir's publishing business in Russia was deeply in debt. There were further complications as well. According to Koblitz (1983, p. 87) Vladimir had helped Zhanna Evreinova escape from her parents by putting her in touch with his bookseller V. Ya. Evdokimov, who in turn put her in touch with people who could help her cross the Polish border, under fire from border guards as it turned out. To express her gratitude, Zhanna wrote a letter to Evdokimov, which was of course opened by the censors and resulted in the arrest of Evdokimov. Vladimir had to borrow 3000 rubles from his father-in-law to get his friend out of trouble.

Literally adding insult to injury, Evreinova and Aniuta were extremely rude to Vladimir when they arrived for a visit in Heidelberg. At least such is the account of Julia Lermontova (Shtraikh 1951, p. 384), who says that Evreinova and Aniuta considered intimacy between Vladimir and Sonya improper and severely discouraged it. Evreinova later became a distinguished jurist and for a time (1885–1889) was editor of the *Northern Messenger (Severny Vestnik)*. The effort spent in getting her abroad seems to have paid off in the longer view, but for Vladimir in the short view it was a disaster.

Vladimir's departure to work on a dissertation in geology was the beginning of a new phase in Sonya's academic career. In the fall of 1870 she journeyed to Berlin to seek out her teacher's teacher, Professor Karl Weierstrass.

1.10. Weierstrass

The man Kovalevskaya was hoping would gain her admission to the University of Berlin was one of the giants of mathematics. To do justice to his contributions would require four or five volumes the size of the present book. The following sketch, based on the article by K.-R. Biermann (1976) and the account of Kochina (1981, pp. 55–62) will give enough information to enable the reader to appreciate his influence on Kovalevskaya's mathematics.

Karl Weierstrass was born 31 October 1815 in Ostenfeld (near Münster in Westfalen). He was the eldest of four children, five years older than his brother Peter, eight years older than his sister Clara, and eleven years older than his sister Elise. His father was a well-educated man, a convert to Catholicism, who worked in various capacities as a school teacher and civil servant. His mother died when he was twelve; and, like Pascal, he received a thorough education from his father. In 1834 Weierstrass enrolled in Bonn University to study law, but left in 1838 without having received a degree. He was always fascinated by mathematics and in 1841 passed an examination in Münster to qualify as a lecturer in the Gymnasium. As we shall see in Chapter 2, Weierstrass was already doing profound research at this time, but much of it was unpublished until his collected works began to appear in 1894. After a probationary year in Münster to earn the title of Gymnasium Teacher, Weierstrass served for six years as a teacher in the Catholic Progymnasium in Deutsch-Krone. In 1848 he became a teacher at the Catholic Gymnasium in Braunsberg. During all this time he was doing mathematics, but apparently his only outlet for publication was the Braunsberg school prospectus. Finally in 1854 he published a paper on Abelian functions in Crelle's *Journal für die reine und angewandte Mathematik*. This paper caused a sensation (see Chapter 3 for its context) and the following year Weierstrass was awarded an honorary doctorate by Königsberg University. His career began to blossom at the relatively late age of 41. On Leibniz' birthday (5 July) in 1857 he gave his *Antrittsrede* at the University of Berlin and joined the famous "round table" formed by the distinguished mathematicians Kummer, Borchardt, and Kronecker. The details of this unheard of event—inviting a Gymnasium teacher to become professor at a university—have been fascinatingly told by K.-R. Biermann (1965).

As a mathematician Weierstrass showed a great zeal for stating things clearly in words. Though he fully appreciated the genius of his younger contemporary Riemann, he did not imitate Riemann's geometrical style. After Riemann's death he sometimes pointed out gaps in proofs which Riemann had given. The most famous of these involves Riemann's proof of what is now called the Riemann mapping theorem. Riemann's proof (1851, §16; *Werke*, p. 30) required a function which minimized the integral $\iint (\phi_x^2 + \phi_y^2)dx\,dy$. Weierstrass (1870) pointed out that one could not be sure such a function exists, since there are functions which make the integral $\int_{-1}^{+1} x^2(\phi'(x))^2\,dx$ arbitrarily small and satisfy $\phi(-1) = a$, $\phi(+1) = b$, namely,

$$\phi(x) = \frac{a + b}{2} + \frac{b - a}{2}\frac{\arctan(kx)}{\arctan(k)}$$

for large k. Yet, if $a \neq b$, there is no continuously differentiable function satisfying these boundary conditions which makes the integral zero. Such a function would have to be constant, which is impossible unless $a = b$. Weierstrass' note caused mathematicians to look for alternative proofs of the Riemann mapping theorem. Not until the work of Hilbert (1901) was Riemann's original argument revived on a sound foundation.

In general it can be said that for Weierstrass the important things in mathematics were clarity and systematic development. Like his illustrious predecessor Gauss, he did not consider a mathematical edifice complete until all the scaffolding was removed. In practice this approach meant that a mathematical topic should be developed from first principles chosen so that the proofs flow naturally without the need for multitudinous special tricks. He applied this systematizing zeal to complex analysis, differential equations, and Abelian integrals, and gave definitive presentations of these subjects in his well-attended lectures in Berlin. His publications were considerable in both quantity and quality, although he did not enjoy publishing. Much of his work is known to us only through the lecture notes and publications of his students.

Weierstrass was subject to attacks called brain spasms by the doctors of the time, and he also suffered occasionally from nervous exhaustion, probably exacerbated by the heavy responsibilities and occasional controversies which he had to deal with at the University of Berlin.

1.11.   Kovalevskaya and Weierstrass

The preceding sketch shows how bold a step Kovalevskaya was taking in approaching Weierstrass for help. She was petitioning an overworked 55-year-old bachelor who was still a very creative mathematician. Moreover, Weierstrass was by no means a champion of the emancipation of women. What reason could he possibly have for aiding a 20-year-old woman who had studied only 1 year beyond calculus? Undoubtedly Kovalevskaya came with good recommendations from her teachers in Heidelberg. Another consideration was probably Prussia's success in the war with France, which had taken most of Weierstrass' students out of the country. Finally, it is possible, though this is entirely speculation, that Weierstrass knew, or was told by

Kovalevskaya, of her maternal grandfather Fyodor Fyodorovich Schubert. One of Weierstrass' papers (1861; *Werke* I, pp. 257–266) begins:

> The Russian general Schubert, in a work which appeared two years ago, has again compared the results of the various measurements to determine the shape of the earth and come to the conclusion that the earth must be regarded as an ellipsoid with three unequal axes.

Although Weierstrass was forbidden by university rules to allow Kovalevskaya to attend his lectures, he did give her a list of problems to solve as a sort of entrance examination. Her solutions were so impressive that he began giving her private lessons twice a week. At the time he apparently gave no thought to any degree for Kovalevskaya, believing that as a married woman she could have no use for a degree. Instead he simply taught her what he was teaching his other students. From the list of his courses given in his collected works (*Werke* III, pp. 355–360) we can get some idea of what this course entailed:

Winter 1870–1871	Elliptic Functions
Summer 1871	Recent Synthetic Geometry
	Selected Problems of Geometry and Mechanics Solvable Using Elliptic Functions
Winter 1871–1872	Abelian Functions
Summer 1872	Introduction to Analytic Functions
	Calculus of Variations
Winter 1872–1873	Elliptic Functions
Summer 1873	Elements of Recent Synthetic Geometry
	Selected Problems of Geometry and Mechanics Solvable Using Elliptic Functions
Winter 1874	Abelian Functions

In 1870–1871 Kovalevskaya settled into a routine, sharing an apartment with Julia Lermontova, who had come to Berlin to study chemistry. Kovalevskaya studied most of the time, except for occasional evenings with Weierstrass and his sisters and a few visits from Vladimir, who had gone to Jena to work on his dissertation. Lermontova (Shtraikh 1951) reports that Kovalevskaya was pale and miserable during most of this period. In the spring of 1871, however, the tide of world events washed over Kovalevskaya's personal life and temporarily swept her away.

Her sister Aniuta and Aniuta's common-law husband Victor Jaclard were in Paris and were fighting as members of the National Guard, which had not accepted the peace treaty which the National Assembly had negotiated with the Prussians. Paris was surrounded by the Prussian army, and the commune which the National Guard had created was soon to be attacked by a French army as well. Sonya and Vladimir set out for Paris, hoping to be allowed through the German lines under the pretext of studying fossil exhibits for Vladimir's dissertation. The ruse was unsuccessful, and they were forced to sneak in using an abandoned rowboat which they had found.

From 4 April to 12 May 1871—more than half the life of the Paris Commune—they lived and worked among the rebels, Sonya in the capacity of nurse, Vladimir actually doing what he pretended to have come to Paris to do, studying fossil collections. They left the Commune when it looked like it would hold out for quite a while, but had hardly returned to Berlin when the news came that Paris had fallen to Thiers' forces. Both Aniuta and Victor were in danger. On several occasions men whom the government believed to be Victor were summarily executed. However, by the time he was finally taken the government had decided to hold public trials. Aniuta was also pursued by the law, but she managed to escape to London, where Karl Marx helped her to find temporary lodging (Koblitz 1983, p. 106). Sonya and Vladimir sought some way to save Victor. Fortunately Sonya's parents also had come to Paris out of concern for Aniuta. General Krukovsky was acquainted with Thiers and managed to persuade him to look the other way when Jaclard's escape was arranged. Jaclard fled to Switzerland using Vladimir's passport. There he and Aniuta were formally married in the presence of Aniuta's parents.

No other period of Kovalevskaya's event-filled life was quite so full of excitement as the spring of 1871. It is a pity that she never carried out her intention to write her recollections of this period.

Back in Berlin in the fall of 1871 Kovalevskaya resumed her ascetic student life, once again with no clear prospect of a degree. An extensive correspondence with Weierstrass, of which only Weierstrass' letters to Kovalevskaya survive (Kochina 1973), begins at this time. [According to Mittag-Leffler (1923c, p. 134) Weierstrass burned all of Kovalevskaya's letters to him when he learned of her death.]

The first 18 months of this correspondence (6 notes) is concerned mainly with administrative matters and arranging appointments. The only explicit mathematics occurs in a letter of 14 January 1872. This letter contains a portion of Weierstrass' development of the algebraic background for his lectures on Abelian integrals, namely, the simplest rational functions on a structure (Gebilde) having prescribed poles. Weierstrass says in the note that he needs the paper back so that he can use it in his next lecture. It is natural to infer that Kovalevskaya had drawn even with Weierstrass' other students by this time, despite her poorer preparation.

Although she had managed to catch up with her coevals intellectually, Kovalevskaya was still laboring under severe handicaps. The worst of these was that she was not allowed to matriculate at the University of Berlin. In addition her relations with Vladimir were again strained. In the fall of 1872 she finally confided to Weierstrass the secret of her fictitious marriage. Although Weierstrass was a conservative gentleman and a faithful Catholic who must have disapproved of such an arrangement, he felt very keenly the anguish Kovalevskaya was suffering. His letter of 26 October 1872 (Kochina 1973, p. 13) is an admirable combination of gentlemanly reticence and warm human concern:

> . . . . .I have been very preoccupied with you this night, and indeed how could it have been otherwise. Although my thoughts have run in the most varied directions, they keep returning to one point, which I must discuss with you today. Do not fear that I will mention things which we have agreed not to talk about, at least right now. What I have to say to you is much more closely related to your scientific efforts. However, I am not sure whether, with the admirable modesty with which you judge what you can now achieve, you will be inclined to give my plan a try. Still all this is better discussed face to face. Therefore even though only a few hours have passed since our last meeting, which brought us so close to each other, please be so kind as to come and talk with me for a while this morning.

Obviously Weierstrass intended to set Kovalevskaya to work on a dissertation. That is in fact what he did, and the results will furnish the subject matter for our next three chapters. Before taking up this subject, however, it may be of interest to discuss the correspondence between Kovalevskaya and Weierstrass in more detail.

The letter just quoted is the last one in which Weierstrass addressed Kovalevskaya by the formal term "Sie." Starting with his next note (4 November 1872) he always addressed her as "Du." About 35 notes and letters from Weierstrass to Kovalevskaya have been preserved from the time when Kovalevskaya worked on her dissertation (November 1872 to August 1874). Since many of these were letters written during holidays, it is not surprising that few of them mention the work she was doing. Several of the early ones are concerned with identities involving theta functions, evidently preparatory to the exposition of the Jacobi inversion problem (see Chapter 3). Quite a few throughout the period involve calculus of variations, especially minimal surfaces. This is a topic which is not touched on in any of Kovalevskaya's published work. Yet she must have known this area very well, since Hermann Amandus Schwarz once wrote to her asking for her help on a minimal-surface problem (Kochina 1980).

The only notes which are definitely related to Kovalevskaya's dissertation are quite late (6 May, 9 May, and 4 July 1874). These contain some detailed instructions on how to write up the results. Although in general these letters do not contain any mathematical revelations, they do give a glimpse of the human relationship between the middle-aged bachelor and his young "married" pupil. As it is difficult to convey these feelings in translation–English, for instance, has no good translation of "liebste"–it may be worthwhile to quote one passage in the original German. It is from a letter of 25 April 1873 (Kochina 1973):

... Du glaubst, wenn nicht die Freundin, so könnte doch die Schülerin mir lästig werden—so lautete das häßliche Wort, das Du brauchst. Daß Du vor der Freundin wenigstens dies nicht befürchtest, könnte ich, wenn ich boshaft wäre, in einem Sinne deuten, gegen den Du vielleicht lebhaft protestieren würdest—aber, in allem Ernst gesprochen, liebste, teurste Sonia, sei versichert, ich werde nie vergessen, daß es die Dankbarkeit meiner Schülerin ist, der ich den Besitz—nicht meiner besten, sondern meiner einzigen wirklichen Freundin verdanke. . ..

An approximately literal translation of this passage follows. Note, however, that in the last line Weierstrass is not calling Kovalevskaya his only real friend, but rather his only real "Freundin," a word for which "woman friend" and "girl friend" are completely inadequate translations.

... You believe that, if not the friend, at least the pupil could become burdensome to me—thus sounds the ugly word you use. If I were malicious, I might construe the fact that you at least have no fear of this for the friend in a sense which you would perhaps vigorously protest—however, in all seriousness, beloved, dearest Sonia, be assured, I shall never forget that it is the gratitude of my pupil to which I owe the possession of, not my best, but my only real friend. . . .

As it happened this letter arrived just in time to prevent an unfortunate misunderstanding on Kovalevskaya's part. She was in Zürich attending a family reunion occasioned by the birth of a son to Aniuta and Victor Jaclard. While there she either made or renewed the acquaintance of Hermann Amandus Schwarz. She also met Elizaveta Fyodorovna Litvinova, another Russian woman who had studied mathematics with Kovalevskaya's old tutor Strannoliubsky and had come to Switzerland to continue her studies. Litvinova wrote in her reminiscences (cf. Kochina 1981, p. 66) that, learning of Weierstrass' appointment as rector of the University of Berlin, Kovalevskaya feared that he would have no more time for her and began making plans to continue her studies with Schwarz.

In a period of 18 months Kovalevskaya wrote three dissertations under Weierstrass' direction, two of which Weierstrass considered respectable enough for any university, and one of which was so outstanding that it left no doubt that Kovalevskaya deserved the doctoral degree. According to Litvinova (1894) Kovalevskaya actually wrote a fourth paper, but withdrew it when a similar paper by Schwarz appeared. In seeking a degree for Kovalevskaya Weierstrass turned to the somewhat more liberal University of Göttingen, where, it will be recalled, Kovalevskaya's great-grandfather had studied. Through the good offices of his friend Fuchs, Weierstrass arranged for Kovalevskaya to be awarded the degree without the customary oral examination. (He did not think her German was good enough, and he feared she might get "stage fright" being questioned by a group of senior scholars. After all, she had never taken any university examinations previously.) Göttingen agreed, and indeed awarded the degree summa cum laude in the fall of 1874, evidently on the recommendation of Fuchs and (Heinrich) Weber.

Kovalevskaya's dissertation received high praise from those who knew of it. Hermann Amandus Schwarz expressed the wish that Fuchs and Koenigsberger could express themselves as clearly as Kovalevskaya. (A typed copy of this letter from Schwarz to Weierstrass, dated 2 October 1874, is in the Institut Mittag-Leffler. I do not know if it has been published.) Hermite, who received a copy of the paper on partial differential equations from Weierstrass under circumstances which will be described in the next chapter, also formed a high opinion of it.

Thus, covered with honors, but exhausted by her exertions, Kovalevskaya achieved the first of her lifetime goals and returned to Russia. The remainder of the first part of this book is devoted to an analysis of these achievements.

CHAPTER 2

Partial Differential Equations

2.1. Introduction

Since the chronology of Kovalevskaya's work with Weierstrass cannot be definitely established, the three papers she wrote during this period will be discussed in the order of their publication. Very likely the topic of the present chapter was the last of the three to be completed, since the result was so significant that it would constitute a dissertation all by itself. To appreciate the significance of this result it is necessary to understand aspects of the development of differential equations in the nineteenth century. The latter subject is vast, and the present account can sketch only part of its growth along one of its many lines of development. Nevertheless the developments which will be discussed were of central importance, being direct consequences of one major idea–regarding the solution of a differential equation as an analytic function of a complex variable.

The organization of the chapter is as follows. First comes a general and elementary discussion of the use of complex variables to solve differential equations (Section 2). Then follows an historical discussion, beginning with the convenient year 1842, in which both Cauchy and Weierstrass wrote important papers (Sections 3 and 4). A quick glance at some of the ensuing work of Briot and Bouquet brings us to the years 1872–1874, when Weierstrass posed a conjecture to Kovalevskaya (Section 5). Kovalevskaya's counterexample to this conjecture and her discovery of the positive assertion which can be made form the main part of the chapter (Sections 6–7). Sections 8 and 9 contain evaluation and ancillary material.

2.2. Differential Equations in the Complex Domain

Most people are aware that imaginary numbers are useful in studying algebraic equations, and it is well known that the systematic study of algebraic equations requires complex numbers, since only in the domain of complex numbers can it be proved that a solution exists for any polynomial equation with real coefficients. Complex numbers play an equally important, but more subtle, role in the study of differential equations. The mere existence of complex numbers is of some use in differential equations. For example, as we shall see in Chapter 4, Laplace gave the general solution of the partial differential equation

$$\frac{\partial^2 u}{\partial x^2} + \frac{\partial^2 u}{\partial y^2} = 0$$

in the form $u = f(x + y\sqrt{-1}) + g(x - y\sqrt{-1})$. (I have been told that this form of the solution may have been given earlier by Euler.) Since the solution of a differential equation is a family of functions rather than a number, the usefulness of complex numbers in differential equations remained limited to special cases of this type until Cauchy developed the theory of analytic functions of a complex variable in the early nineteenth century. Once this theory was developed, it opened up new possibilities through the systematic (as opposed to the ad hoc) use of power series as well as characterizing functions by their singularities. Although the two lines of development cross in many places, it will be useful to separate them for purposes of the present discussion, concentrating on the use of power series. Before proceeding to a systematic discussion it may be well to remark that complex variables, though useful mathematically, seem somewhat unnatural in studying the differential equations which arise in physics, since the variables represent physical quantities (time, for instance) for which imaginary values had no apparent meaning in the nineteenth century.

Eighteenth-century mathematicians had often used power series to represent functions for which no simpler expression could be found. Commonly the series represented some physical quantity for which a differential equation was known. It was assumed because of the physical interpretation that such a function existed. (Probably if you presented some sophisticated modern counterexample to an eighteenth-century mathematician, he would conclude only that he had the wrong equation, not that no such physical quantity could exist.) The acceptance of power series as functions greatly enlarged the class of solvable differential equations and led to a general technique of wide applicability, nowadays known as the method of undetermined coefficients. For example, to solve the equation $x\, dy/dx = \sin x$, whose solution cannot be expressed using only a finite number of elementary functions, one can use the Taylor series

$$\sin x = x - \frac{x^3}{3!} + \frac{x^5}{5!} - \frac{x^7}{7!} + \cdots,$$

cancel x on both sides, and write the equation as

$$\frac{dy}{dx} = 1 - \frac{x^2}{3!} + \cdots$$

whence the solution is

$$y = C + x - \frac{x^3}{18} + \cdots.$$

The general technique which the preceding example illustrates works as follows, for an ordinary differential equation. Assume $y = c_0 + c_1 x + c_2 x^2 + \cdots$ and use the equation itself to calculate the coefficients c_j. For example, the equation

$$\frac{dy}{dx} = 1 + xy^2 \tag{1}$$

leads to the system of equations

$$
\begin{aligned}
c_1 &= 1, \\
2c_2 &= c_0^2, \\
3c_3 &= 2c_0 c_1, \\
&\vdots
\end{aligned}
\tag{1$'$}
$$

$$nc_n = c_0 c_{n-2} + c_1 c_{n-3} + \cdots + c_{n-2} c_0.$$

This technique adds a layer of abstraction to what already existed. Besides accepting a function as defined when its Taylor coefficients are known, we now have to accept the coefficients as known even though we may not be able to calculate them explicitly. In the example of system $(1')$ it is obvious that c_0 can be completely arbitrary; but once c_0 is chosen, the remaining coefficients are uniquely determined.

Eighteenth-century mathematicians used the method of undetermined coefficients in a formal way, though some of them were nervous about the validity of such manipulations when applied to imaginary numbers. The systematic study of convergence questions arose with Cauchy. Abel, who had studied Cauchy's *Cours d'Analyse* carefully, took the trouble to discuss the uniformity of the convergence of the binomial series

$$1 + \frac{m}{1}x + \frac{m(m-1)}{1 \cdot 2}x^2 + \cdots$$

for complex values of x, finding necessarily complicated expressions for the real and imaginary parts of the sum (Abel 1826).

Once it is recognized that convergence might be a problem, one must worry about the soundness of the method of undetermined coefficients. For example, even though we know that the system $(1')$ generates a unique set of coefficients for given c_0, it is not immediately clear—though it is not hard to prove—that the power series so generated has positive radius of convergence. This problem gave rise to papers by Cauchy and Weierstrass, and eventually by Kovalevskaya. Let us now look at these papers.

2.3. The Year 1842: Cauchy

The first person to worry in print about convergence questions involved with the method of undetermined coefficients was Cauchy. In the first in a series of papers (1842a) published in the *Comptes Rendus* he gave a summary of the problem as he saw it. This summary gives such a clear statement of the problem that it is worthwhile to quote it in full:

In the theory of equations mathematicians have properly considered fundamental the question whether every equation has a root. Similarly in the integral
calculus one of the most important questions, a fundamental question, is obviously whether every ordinary or partial differential equation can be integrated.
But—and this ought to surprise us at first sight—despite the numerous works
of mathematicians on the integral calculus, this question, important though it
be, is nowhere solved in full generality. To be sure the existence of general
integrals of ordinary differential equations, which contain only one independent
variable, is now established by two different methods which I have given, the
first in my lectures at the Ecole Polytechnique, the second in a lithographed
memoir of 1835. [According to Osgood (1901, p. 37) this memoir can be found
in Cauchy's *Exercices d'analyse* (1841, p. 355).] In addition the existence of
general integrals of partial differential equations is established in certain cases
where one is able to integrate these equations, for example when the equation
reduces either to a single equation of first order or to linear equations in which
the coefficients of the unknowns and of their derivatives remain constant. But
does an arbitrary system of ordinary or partial differential equations always
admit a corresponding system of general integrals? Such is the problem which
seemed to me worthy of the attention of mathematicians. The present solution
is based on considerations which I shall explain briefly.

For a long time mathematicians, supposing without proof that every
ordinary or partial differential equation admits a general integral, have considered Taylor's formula as the means of developing this integral in a series of
increasing integer powers of an increment i given to an independent variable t,
which can be considered as representing time. Further, using a theorem which
I proved in 1831 relating to the development of functions, one can be sure that
in the case where the series so obtained is convergent, the sum of the series
satisfies, as an integral, the ordinary or partial differential equation, at least for
real or complex values of the increment i whose moduli do not exceed a fixed
bound. Moreover the same remark applies to the sums of the series obtained
when, assuming the existence of general integrals of a system of ordinary or
partial differential equations, one sets about developing them in Taylor series.
But in all cases it remains to be proved that the series so obtained is convergent,
at least for i of sufficiently small modulus. Now this end can be achieved using
a fundamental theorem which not only determines a bound beneath which the
modulus of i may vary arbitrarily without causing the series obtained to diverge,
but also determines a bound on the error caused by terminating each series after
a certain number of terms. The proof of this theorem is based, as will subsequently be seen, on the principles of the new calculus which I have called the
calculus of bounds ["calcul des limites"] and on a device of analysis which can
be given many useful applications.

From this passage one can infer an analogy in Cauchy's mind between
the role played in algebraic equations by complex numbers and the role he
expected analytic functions to play in differential equations. He considered it
his task to prove the analog of the fundamental theorem of algebra, i.e., to
prove that every differential equation with analytic coefficients has an analytic
solution. The purpose of the series of papers he was about to write was to
develop this idea.

Cauchy's calculus of bounds, or as it is now called in English, method
of majorants, is based on two simple observations, which can be distilled

from the discussion which follows the passage just quoted. Cauchy observed that if $u = f(x, y, z, \ldots, t)$ is a function which is analytic when the variables are given small increments $\bar{x}, \bar{y}, \bar{z}, \ldots, \bar{t}$ whose moduli he denoted by x, y, $z, \ldots, t$, and V is the maximum value of $f(x + \bar{x}, z + \bar{z}, \ldots, t + \bar{t})$ when $\bar{x}, \bar{y}, \bar{z}, \ldots, \bar{t}$ are held fixed, then the derivatives must satisfy

$$\text{mod. } D_x^l D_y^m \cdots D_t^n f(x, y, z, \ldots, t) < N \frac{V}{x^l y^m \cdots t^n},$$

the value of N being $l! \, m! \cdots n!$. This inequality is the basic principle of the method of majorants. The "device of analysis" to which Cauchy referred may have been this principle, or it may have been the particular case which occurs when one takes $f(x, y, z, \ldots, t) = ax^{-1}y^{-1} z^{-1} \cdots t^{-1}$ with a constant. For this function the inequality just written is replaced with equality when x is replaced by $(-x)$, y by $(-y)$, etc., V by $f(x, y, \ldots, t)$, and mod is removed.

Cauchy was then able to justify the method of undetermined coefficients for a large class of equations, namely, those for which the series so generated has coefficients which "reduce to polynomials made up of terms each of which is the product of a [natural] number or more generally a positive constant and the derivatives of various orders of several functions u, v, w, ... or even powers of these derivatives." For such equations, if the series generated by the method of undetermined coefficients is $I_0 + I_1 i + I_2 i^2 + \cdots$, Cauchy was able to construct a majorant series $\mathcal{I}_0 + \mathcal{I}_1 i + \mathcal{I}_2 i^2 + \cdots$ such that $|I_j| < \mathcal{I}_j$ for all j and prove that the latter series converges. For, as he said, "it suffices to integrate the equations in the special case when each of the functions which make up their second members becomes inversely proportional to the variables on which it depends."

A digression seems justified at this point to clarify what Cauchy meant. (A full discussion of the case of ordinary equations is given in Appendix 1.) Consider an ordinary differential equation

$$\frac{dy}{dx} = f(x, y) = \sum_{k=0}^{\infty} \sum_{j=0}^{\infty} a_{jk} x^j y^k.$$

The method of undetermined coefficients proceeds by assuming an expansion $y = \sum_{n=0}^{\infty} b_n x^n$ and solving the resulting system of algebraic equations for b_n. The problem is to show that the resulting series converges. To this end, note that the differential equation requires three operations on the power series, namely, differentiation, raising to power, and substituting one series in another. As far as the coefficients are concerned these operations involve either multiplying them by a non-negative integer or multiplying them by one another and adding them. Hence the operations preserve non-negativity of coefficients. It follows that if you replace $f(x, y)$ by a function $F(x, y) = \sum_{k=0}^{\infty} \sum_{j=0}^{\infty} A_{jk} x^j y^k$ such that $|a_{jk}| < A_{jk}$ for all k and take B_0 larger than $|b_0|$, you will generate a formal solution $Y = \sum_{n=0}^{\infty} B_n x^n$ with $|b_n| \leq B_n$ for all n.

Since you can choose $F(x, y)$ to suit yourself, you may be able to arrange that the series for Y converges. It then follows that the series for y converges, which was the original goal. The method really does work, and various choices for the majorant $F(x, y)$ are possible. The digression now being finished, let us resume the discussion of Cauchy's papers.

Cauchy applied these principles in a subsequent paper (1842c) to study what he called a linear partial differential equation of first order, $D_t \bar{\omega} = A D_x \bar{\omega} + B D_y \bar{\omega} + \cdots + K$. (This equation is actually what we now call quasilinear, since the coefficients $A, B, \ldots, K$ were assumed to depend on $\bar{\omega}$ as well as $x, y, z, \ldots, t$.) Cauchy obtained the solution of this equation with the initial conditions $\bar{\omega} = \omega$, $D_x \bar{\omega} = 0$, $D_y \bar{\omega} = 0, \ldots$ at the point $t = \tau$ in the form of the relation

$$\log\left(\frac{t}{\tau}\right) = \int_0^{\bar{\omega} - \omega} (\bar{\omega} - \theta)\left(x + a\frac{\theta}{k}\right)\left(y + b\frac{\theta}{k}\right) \cdots \frac{1}{k} \, d\theta.$$

When the coefficients are taken as $A = ax^{-1}y^{-1}z^{-1} \cdots t^{-1}, \ldots, K = kx^{-1}y^{-1} \cdots t^{-1}$. The derivatives of these functions, as shown above, must majorize those of the original coefficients, thus enabling one to prove convergence of the power series.

Cauchy had thus shown that a (quasi)linear equation has an analytic solution. He then remarked that a general first-order equation [evidently something like $f(x, y, t, \bar{\omega}, D_x \bar{\omega}, D_y \bar{\omega}, D_t \bar{\omega}) = 0$] can be reduced to the quasilinear case by differentiating with respect to the independent variables. One then obtains a new system of equations which is quasilinear in the partial derivatives $D_x \bar{\omega}$, $D_y \bar{\omega}$, $D_t \bar{\omega}$.

In concluding this summary of Cauchy's work it should be emphasized that he did not mention "uniqueness." He may have taken for granted that the solution is determined by the initial values. He does say in one place that the Taylor coefficients can be easily calculated, but he does not give details or discuss the role of the initial conditions in the calculation. Since the problem of determining the solution from the equation and the initial conditions is nowadays known as the Cauchy problem, one may well ask where Cauchy formulated this problem. Freudenthal (1971) suggests Cauchy's 1815 paper on waves as a possibility. An unpublished analysis of this paper by Grattan-Guinness (1982, 9.3.5) brings out the fact that the initial conditions played an essential role. In a private communication Grattan-Guinness pointed out to me that the use of initial conditions to solve problems was not new in 1815, and that, of course, the Cauchy problem as nowadays understood applies to *general* partial differential equations.

Finally it should be remarked that Cauchy's work was all "local." That is, he was satisfied to have proved that the solution is given by a power series with positive radius of convergence. He did not worry about the exact radius of convergence or the continuation of the solution outside the circle of convergence.

2.4. The Year 1842: Weierstrass

By coincidence Weierstrass also was occupied by problems similar to those
which Cauchy studied and indeed he proved some basic theorems by a similar
method in the year 1842. Since he did not publish this work until 1894 and
even then published only part of it, his achievements in this area were
communicated mostly to his students and colleagues in Berlin after 1857.
Since his published work during this time was mostly on Abelian integrals and
analytic function theory, one might wonder how he came to be working on
differential equations as well. The answer, which it is hoped will be borne out
by the discussion below, is that he was interested in showing that a differential
equation could be used as the definition of an analytic function. Hence his
work in this area is indeed part of his work on analytic function theory (see
also Manning, 1975).

In the published version of the work on differential equations (*Werke* I:
pp. 75–84) Weierstrass considered a system of ordinary differential equations

$$\frac{dx_j}{dt} = G_j(x_1, \ldots, x_n), \qquad j = 1, 2, \ldots, n. \tag{2}$$

Here the functions G_j are polynomials. Weierstrass proved by a method which
is in most details the method of majorants described above that $x_1, \ldots, x_n$
are analytic functions of t, using as a majorant the function $(1 + x_1 +
\cdots + x_n)^c$. He also remarked that the coefficients of the polynomials G_j
could be analytic functions of other variables $u_1, \ldots, u_m$, in which case the
solution is of the form

$$x_j(u_1, \ldots, u_m; t) = \sum_{k=0}^{\infty} A_k(u_1, \ldots, u_m) t^k$$

and is an analytic function of all the variables. Two facts were emphasized
by Weierstrass. First, the solution is analytic no matter what the initial values
are; second, the solution is uniquely determined by the initial values. This
second point, as noted, was not stressed by Cauchy. The reason Weierstrass
emphasized it is clear from sections 2 and 3 of the paper. These sections
contain a discussion of analytic continuation by power series, in essentially
the form taught in textbooks today. It goes as follows.

Denoting the unique solution of the system (2) which satisfies the initial
conditions $x_j = a_j$ when $t = t_0$ by $x_j = B_j(t - t_0; a_1, \ldots, a_n)$ and assuming
$C_j^{(k)}$ are the Taylor coefficients of B_j, so that all the series $x_j = a_j + \sum_{k=1}^{\infty} C_j^{(k)}$
$(t - t_0)^k$ converge when $|t - t_0| < r$, say, suppose that t_1 is such that
$|t_1 - t_0| < r$. Let $a_j' = B_j(t_1 - t_0; a_1, \ldots, a_n)$ for values of t_1 near t_0. The
uniqueness of the solutions implies that

$$B_j(t - t_0; a_1, \ldots, a_n) = B_j(t - t_1; a_1', \ldots, a_n'). \tag{3}$$

Weierstrass denoted the common region of convergence of the series on the left-hand side of this last equation by T_0 and the common region of convergence for the series on the right by T_1. To continue in his own words,

> Now three cases can occur:
> The region T_0 can be unbounded. Then the same is true of the region T_1 and [3] holds for all vaues of t_1 . . . .
> If, however, the region T_0 is bounded, and hence geometrically represented as the inside of a circle described about the point t_0 (let its radius be denoted ρ_0), then the region T_1 is also bounded and is represented by the inside of a circle described about t_1 whose radius (ρ_1) lies in the interval $\rho_0 - \delta$ . . . $\rho_0 + \delta$, where $\delta = |t_1 - t_0|$. It then follows that the quantity ρ_1 varies continuously, never reaching the value 0 however, when the point t_1 changes its position in a continuous manner inside T_0 . . . .
> Finally it can happen that the quantity ρ_1 assumes an infinitely small value for every point t_1 which approaches arbitrarily close to the boundary of T_0 . . . .
> This being established, one can now produce a series of arbitrarily many systems
>
> $$B_1(t - t_0; a_1, \ldots, a_n), \ldots, \quad B_n(t - t_0; a_1, \ldots, a_n)$$
> $$B_1(t - t_1; a_1', \ldots, a_n'), \ldots, \quad B_n(t - t_1; a_1', \ldots, a_n')$$
> $$B_1(t - t_2; a_1'', \ldots, a_n''), \ldots, \quad B_n(t - t_2; a_1'', \ldots, a_n'')$$
>
> etc. in many different ways, so that each one after the first is constructed from its predecessor by following the procedure just given for constructing the second from the first. Then by the totality of these systems a (single- or multiple-)valued system of analytic functions of the variable t is defined.

Weierstrass' point of view is clear from the passage just quoted. He wishes to use the differential equation as the definition of the function. He therefore undertook to prove the existence and uniqueness of the solution in such a way as to make the construction compatible with analytic continuation.

2.5.   The Problem Posed to Kovalevskaya

Cauchy's theorem gives information about the solutions of a differential equation around "ordinary" points where the solutions are analytic. Subsequently Briot and Bouquet (1856) studied the solutions around singular points. Because of the representation theorems which show how to determine an analytic function from its singularities, these points turn out to be important for the study of differential equations. Such a study was undertaken by Fuchs (1865). Fuchs introduced the distinction between fixed singularities, which do not depend on the initial conditions, and movable singularities, which do. He noted that linear equations have only fixed singularities and (1884) found conditions for nonlinear equations to share this property (cf. Gray 1984).

More in the spirit of Weierstrass' work was a paper of Jacobi, published posthumously in 1865, which showed how to reduce a general algebraic differential equation to the special form

$$G(x, y_0, \ldots, y_n) = 0,$$ (4)

$$D_2 G(x, y_0, \ldots, y_n) \frac{dy_j}{dx} - G_j(x, y_0, \ldots, y_n) = 0.$$

Here y_0 is an auxiliary variable defined by the first of equations (4) and G, $\ldots$, G_n are polynomials. The form (4) does not explicitly occur in Jacobi's 1865 paper, but Jacobi does give a definition of a "non-canonical" form of equation which has to be transformed to "canonical" form. It appears that the term "normal form" which Kovalevskaya used was first used by Jacobi (*Werke* 5; pp. 483–513).

Since Weierstrass' theorems apply to ordinary differential equations, it was clear to him that the extension to partial differential equations needed to be done. Weierstrass knew of the work of Briot and Bouquet, but did not consider it adequate. It is somewhat surprising, considering that Weierstrass did know of the work of Briot and Bouquet, that he apparently did not know of Cauchy's work. The explanation seems to be that Briot and Bouquet give no explicit reference to Cauchy's work, although they mention him constantly. Weierstrass published an extension to partial differential equations in a paper on analytic faculties (1856a, pp. 43–44) which, however, was omitted when the paper was reproduced in his collected works. Apparently he thought that a general theorem could be proved to the effect that a power series obtained formally from a partial differential equation in which only analytic functions occur would necessarily converge. He says in a letter of 25 September 1874 to du Bois-Reymond that he had made such a conjecture (Mittag-Leffler 1923d, p. 204). It is this conjecture which Kovalevskaya was evidently supposed to prove, if possible.

2.6. Kovalevskaya's Paper

In the introduction to her memoir (1875) Kovalevskaya quotes a theorem which she says she derived from Weierstrass' lectures in connection with the system (4) above. The theorem states that if a, b_{00}, b_{10}, $\ldots$, b_{n0} are chosen so that $G(a, b_{00}, b_{10}, \ldots, b_{n0}) = 0$ and $D_2 G(a, b_{00}, b_{10}, \ldots, b_{n0}) \neq 0$, then constants b_{jk}, $k = 0, 1, 2, \ldots$; $j = 1, 2, \ldots$, can be chosen in precisely one way so that the series

$$y_j = \sum_{k=0}^{\infty} b_{jk} \frac{(x - a)^k}{k!}$$

formally satisfy (4). Moreover all these series have positive radius of convergence, and hence represent analytic functions. Every branch of this system of analytic functions satisfies the given system of equations.

Kovalevskaya worked on a quasilinear system of partial differential equations whose form was similar to that given by Jacobi, namely,

$$\frac{\partial \phi_\gamma}{\partial x} = \sum_{\alpha=1}^{r} \sum_{\beta=1}^{n} G_{\alpha\beta}^{(\gamma)}(\phi_1, \ldots, \phi_n) \frac{\partial \phi_\beta}{\partial x_\alpha}, \qquad \gamma = 1, 2, \ldots, n. \qquad (5)$$

The coefficients $G_{\alpha\beta}^{(\gamma)}$ are assumed analytic around $(0, \ldots, 0)$. Using the majorant technique of Cauchy and Weierstrass she proved the following theorem:

A. If $\phi(x_1, \ldots, x_r)_{10}, \ldots, \phi(x_1, \ldots, x_r)_{n0}$ are n arbitrarily-chosen power series having a common region of convergence and if all of them are zero when $x_1 = 0, x_2 = 0, \ldots, x_r = 0$, then there are determined n power series in $(x, x_1, \ldots, x_r)$ which for $x = 0$ become $\phi(x_1, \ldots, x_r)_{10}, \ldots, \phi(x_1, \ldots, x_r)_{n0}$, respectively, and which formally satisfy (5) when put in place of $\phi_1, \ldots, \phi_n$.
B. All n of these series converge unconditionally in some region and represent functions which actually satisfy (5).

In addendum to the first section of her paper Kovalevskaya pointed out several extensions of this theorem. The most important of these, the one on which the proof of her major theorem is based, is the following. For a system of partial differential equations of the form

$$G^{(\gamma)}(\phi_1, \ldots, \phi_n) \frac{\partial \phi_\gamma}{\partial x} = \sum_{\alpha=1}^{r} \sum_{\beta=1}^{n} G_{\alpha\beta}^{(\gamma)}(\phi_1, \ldots, \phi_n) \frac{\partial \phi_\beta}{\partial x_\alpha}$$
$$+ G_{00}^{(\gamma)}(\phi_1, \ldots, \phi_n), \qquad (6)$$

$\gamma = 1, 2, \ldots, n$, in which all $G^{(\gamma)}$ and $G_{\alpha\beta}^{(\gamma)}$ are power series in $\phi_1, \ldots, \phi_n$, a set of n power series $\phi_1(x, x_1, \ldots, x_r), \ldots, \phi_n(x, x_1, \ldots, x_r)$ which formally satisfies (6), in the sense that the coefficients in the power series satisfy the algebraic equations which result when the method of undetermined coefficients is applied, will have a positive radius of convergence, i.e., all the series will converge in some region, provided certain conditions are met, namely, (a) the series $\phi_1(0, x_1, \ldots, x_r), \ldots, \phi_n(0, x_1, \ldots, x_r)$ all converge in some region; (b) the set of values $\phi_1(0, 0, \ldots, 0), \ldots, \phi_n(0, 0, \ldots, 0)$ lies in the region of analyticity of the functions $G_{\alpha\beta}^{(\gamma)}$ and $G^{(\gamma)}$; (c) none of the functions $G^{(\gamma)}$ vanishes at the point $\phi_1 = \phi_1(0, 0, \ldots, 0), \ldots, \phi_n = \phi_n(0, 0, \ldots, 0)$.
The method Kovalevskaya used to prove this extension is once again the method of majorants, i.e., replacing the original system by the system

$$\frac{\partial \psi_\gamma}{\partial x} = \sum_{\alpha=1}^{r} \sum_{\beta=1}^{n} \overline{G}_{\alpha\beta}^{(\gamma)}(\psi_1, \ldots, \psi_n) \frac{\partial \psi_\beta}{\partial x_\alpha}, \qquad \gamma = 1, 2, \ldots, n,$$

with $\overline{G}_{\alpha\beta}^{(\gamma)}$ taken as a majorant of $G_{\alpha\beta}^{(\gamma)}$ of the form

$$\overline{G}_{\alpha\beta}^{(\gamma)}(\psi_1, \ldots, \psi_n) = \frac{G}{1 - \dfrac{\psi_1 + \cdots + \psi_n}{g}}.$$

By using this variant of Weierstrass' $(1 + x_1 + \cdots + x_n)^c$, Kovalevskaya was able to extend the proof from the Jacobi case to the more general system (6).

In the second section of her paper Kovalevskaya introduced the concept which is the key to extending these results to general systems of partial differential equations. This concept, the case which she called "normal," involves a partial differential equation

$$G\left(x, x_1, \ldots, x_r, \phi, \ldots, \frac{\partial^{\alpha + \alpha_1 + \cdots + \alpha_r}\phi}{\partial x^\alpha \partial x_1^{\alpha_1} \cdots \partial x_r^{\alpha_r}}, \ldots\right) = 0, \qquad (7)$$

where G is a polynomial. If the highest-order partial derivative which occurs in G is n, Kovalevskaya called the equation "normal" if some pure derivative of order n actually occurs in G. The phrase "actually occurs" (Kovalevskaya's words) seems to mean that equation (7) can be *solved* for that derivative, i.e., that the derivative of G with respect to the variable replaced by the pure derivative of order n is not identically zero. Her notation was chosen so that the variable whose pure derivative of order n occurs is denoted x, and the remaining variables by $x_1, \ldots, x_r$. It might have been clearer to say "normal with respect to x," since in general the pure derivative of order n will occur for some, but not all of the variables involved.

After showing that a convergent power series

$$\phi = \sum_{\nu=0}^\infty \overset{(\nu)}{\phi} (x_1, \ldots, x_r | a_1, \ldots, a_r) \frac{(x-a)^\nu}{\nu!}$$

can always be chosen uniquely so as to satisfy equation (7) and that moreover the first n coefficients could be arbitrary analytic functions of $x_1, \ldots, x_r$, she went on to show by combinatorial work that if a function $\phi(x, x_1, \ldots, x_r)$ analytic around $(a, a_1, \ldots, a_r)$ satisfies the equation, then the coefficients of its Taylor expansion around this point can be determined from the differential equation. It is these results which are now commonly called the Cauchy–Kovalevskaya theorem. Note, however, that the hypotheses are somewhat more restricted than might appear, since the function G must be a polynomial. This restriction enables the equation to be reduced to the form considered by Jacobi.

In the third section of her paper Kovalevskaya showed how to bring any equation of the form (7) to normal form by a linear change of variable. At this point it begins to look as if every partial differential equation with analytic coefficients has analytic solutions. The matter is more subtle than appears, however, since the problem one usually faces in practice is a differential equation plus initial conditions, what is called the Cauchy problem. This problem is important for equations of hyperbolic type, the wave equation, for instance. For equations of elliptic type such as Laplace's equation Hadamard (1935) showed that the solution of the Cauchy problem may not depend continuously on the initial data, which is a significant defect in physical applications, where initial data are known only approximately. (For a general discussion of well-posed problems see Hadamard 1923). For a parabolic equation Kovalevskaya showed that the solution may not be analytic if the initial data are imposed on the wrong variable. Her counterexample astounded Weierstrass and led him to undertake new investigations to find solutions to this equation.

Kovalevskaya's counterexample was simple and elegant. The one-dimensional heat equation is

$$\frac{\partial u}{\partial t} = \frac{\partial^2 u}{\partial x^2}.$$

As is well known, $u(x, t)$ represents the temperature at time t at a point on a straight wire x units to the right of a reference point taken as 0. Thus u represents a physical quantity, so that one certainly expects a solution to this equation to exist when an initial temperature distribution $u(x, 0)$ is prescribed. Kovalevskaya showed that the method of undetermined coefficients applied to this problem leads to a divergent series if the initial distribution is $u(x, 0) = (x - 1)^{-1}$. This example, though mathematically impeccable, shows that Kovalevskaya's thought was not really on physical applications, since this temperature distribution is infinite at one point of the wire. Kovalevskaya could have given a physically realizable counterexample just as easily by taking $u(x, 0) = (1 + x^2)^{-1}$. The power series which this problem determines is

$$u(x, t) = \sum_{n=0}^{\infty} \sum_{m=0}^{\infty} (-1)^{m+n} \frac{(2m + 2n)!}{(2m)!n!} x^{2m} t^n.$$

However, when t is anything other than 0 this series diverges. In fact when $x = 0$ it becomes the series

$$u(0, t) = \sum_{n=0}^{\infty} (-1)^n \frac{(2n)!}{n!} t^n$$

whose radius of convergence is clearly zero. The fact is that this problem has no solution which is jointly analytic in x and t around $t = 0$. Its only well-behaved solution is

$$u(x, t) = \int_0^{\infty} e^{-y - ty^2} \cos xy \, dy,$$

which is analytic only when the real part of t is positive.

It may be instructive to see what positive assertion Kovalevskaya's theorem makes about the heat equation. The initial-value problem for this equation is in normal form if the initial conditions are imposed using x instead of t. Thus the Cauchy problem

$$\frac{\partial u}{\partial t} = \frac{\partial^2 u}{\partial x^2},$$

$$u(x_0, t) = f(t),$$

$$\frac{\partial u}{\partial x}(x_0, t) = g(t)$$

has a unique solution which is analytic about (x_0, t_0) if f and g are analytic at t_0. In summary, the solution is analytic when the temperature and temperature gradient at some point of the wire are prescribed as analytic functions of time, but the solution may fail to be analytic when the initial temperature distribution at some instant is an analytic function of position. Kovalevskaya showed, however, that the solution will still be analytic even in this case if the initial temperature distribution is an entire function of position, i.e., has no poles even at points not corresponding to real values of x.

The fourth and final section of the paper contains a generalization of the Cauchy–Kovalevskaya theorem to systems of partial differential equations, i.e., to more than one such equation. Again, this result was achieved by reducing to a system of the form considered by Jacobi.

2.7. Unpublished Work

From Weierstrass' correspondence we know that Kovalevskaya actually did more work than she published. On 21 July 1874 he wrote to Fuchs (Mittag-Leffler 1923, 1923f, pp. 255–256) that she had originally planned an appendix, "Applications and Conclusions," which was left out because of the excessive labor it would have demanded. He mentioned, however, that she had completely independently undertaken to apply her theorem to a famous theorem of Legendre, giving a concise proof that a rotating liquid whose shape is approximately spherical must necessarily have the shape of an ellipsoid of rotation. Her theorem enters the proof in showing that the power series expansion for the potential of such a body is valid at the boundary of the body. This problem, or one like it, had been considered by Weierstrass' student Bruns in his 1870 dissertation. It seems likely, therefore that Kovalevskaya's posthumously published paper (1891) was actually part of her 1874 dissertation. This point will be taken up again in Chapter 8.

2.8. Publishing the Result

Kovalevskaya worked this material into its final form in July 1874 and submitted it to the University of Göttingen as one of three dissertations for the doctoral degree. The following month she submitted it to the *Journal für die reine und angewandte Mathematik*, where it appeared in 1875. While the article was in press, Kovalevskaya returned to Russia. The final episode in the story is told in a letter from Weierstrass to Kovalevskaya dated 21 April 1875

(Kochina 1973, pp. 67–70). Weierstrass says that he received the early 1875 issues of the *Comptes Rendus* rather late, as a consequence of a delay in renewing his subscription. He was amazed to find in them two articles by Darboux "On the existence of an integral in differential equations containing an arbitrary number of functions and independent variables." He continues,

> . . . So you see, my dear, that this question is one which is awaiting an answer, and I am very glad that my student was able to anticipate her rivals in time and at least not fall behind them in working out the problem.
>
> Darboux mentions several exceptional cases which are of special interest; I am inclined to think that he also has encountered the difficulties (as in the equation $\partial \phi / \partial t = \partial^2 \phi / \partial x^2$) which gave you so much trouble at first and which you later overcame so successfully . . . .

Weierstrass sent a copy of Kovalevskaya's dissertation to Darboux's mentor Hermite in Paris and advised Kovalevskaya to ask Borchardt, the editor of the *Journal für die reine und angewandte Mathematik,* to state in the journal that Kovalevskaya's article had been received in August 1874. His fears of a priority dispute were exaggerated, however. Hermite and Darboux became Kovalevskaya's closest friends and admirers in Paris and were later instrumental in promoting her participation in the Bordin Competition for 1888, in which she won international fame and 5,000 francs.

In fact the priority dispute was effectively mooted, as Weierstrass should have noticed, by a letter from A. Genocchi which appeared in No. 3 of the *Comptes Rendus* in 1875, immediately ahead of one of Darboux's papers. Genocchi gave a précis of the papers of Cauchy, saying, "I conclude that, for partial differential equations, as for ordinary differential equations, the first proof of the existence of the integral is due to Cauchy. Undoubtedly the very large number of writings of the celebrated analyst must excuse those who are unfamiliar with all the results obtained by him."

Like most situations in which mathematicians have proved similar results by different methods at nearly the same time, the priority claim is impossible to settle definitively. Darboux (1875) proved the existence of an analytic solution to the initial-value problem

$$\frac{\partial V}{\partial t} = F(p_1, p_2, \ldots, p_n, q_1, \ldots, q_n, t), \qquad \left(p_j = \frac{\partial V}{\partial q_j} \right),$$

$$V(q_1, \ldots, q_n, 0) = f(q_1, \ldots, q_n),$$

where f and F are given analytic functions. This result seems to be contained in Kovalevskaya's, though her explicit arguments are given for equations where F is a polynomial. Cauchy's explicit arguments are given for the quasilinear case only. He does not seem to have discovered any limitations on this sort of theorem. It goes without saying that Kovalevskaya's work was completely independent of the work of Cauchy, even though there is some duplication. The important concept of normal form, which brings order to the whole topic, is due to Kovalevskaya, though probably inspired by Jacobi.

2.9. Evaluation of the Work

Taken in the context of the time, Kovalevskaya's paper can be considered significant for at least three reasons. First, it gave systematic conditions under which the method of undetermined coefficients must work. Second, it charted the terrain, so to speak, for the application of analytic function theory in differential equations, showing under what conditions a differential equation was likely to have analytic solutions. Third, it showed that a differential equation could be used as the definition of an analytic function, when taken together with certain intitial conditions.

The modern evaluation of Kovalevskaya's work must be based on the influence it has had on subsequent work. A good source for such an evaluation is provided by the article of Oleinik (1975). As far as a nonspecialist can judge, the Cauchy–Kovalevskaya theorem seems to play a very basic role in the study of differential equations. Even in the most recent work it is frequently invoked in the proof of new results. A few examples will now be given by way of illustration.

(1) In his 1908 master's thesis S. Bernstein proved that if a function z satisfying

$$F\left(\frac{\partial^2 z}{\partial x^2}, \frac{\partial^2 z}{\partial x\,\partial y}, \frac{\partial^2 z}{\partial y^2}, \frac{\partial z}{\partial x}, \frac{\partial z}{\partial y}, z, x, y\right) = 0$$

with F analytic, has continuous partial derivatives up to third order and $4D_1FD_3F - (D_2F)^2$ is positive in a region, then z is analytic. This theorem represents a considerable weakening of the hypotheses of Kovalevskaya's theorem for the type of equation considered.

(2) In his first mathematical paper Ivar Fredholm (1890) used the fact that Kovalevskaya's solution of the heat equation is not analytic at any point of the plane $t = 0$ to construct an analytic function having the unit circle as its natural boundary, namely $\Sigma_{n=0}^{\infty} a^n z^{n^2}$, which becomes a solution of the heat equation when $a = e^x$, $z = e^t$.

(3) Oleinik (1975, p. 8) points out that many of the theorems which show the necessity of the hypotheses used in proving the Cauchy–Kovalevskaya theorem invoke the theorem itself in their proof.

(4) Not only the theorem, but also the method of majorants used in its proof have found continued application. The method of majorants is used to establish some of the fundamental results in the theory of algebraic functions (Bliss 1933, pp. 17–18). A particular case of the Cauchy–Kovalevskaya theorem needed for the theory of Lie groups was proved by this method in the book of P. M. Cohn (1957, pp. 154–157).

Two more points need to be made before we leave the subject of this paper. The first is the relation between the theorem Kovalevskaya proved and modern formulations of the "Cauchy–Kovalevskaya theorem." The modern form of the theorem was stated by Kochina in an essay reproduced in translation in the book by Stillman (1978, pp. 231–235). Certain misprints were corrected and the notation was abbreviated in this quotation.

. . . Consider the system of equations

$$\frac{\partial^{n_i} u_i}{\partial t^{n_i}} = F_i \left(t, x_1, \ldots, x_n, u_1, \ldots, u_N, \ldots, \frac{\partial^k u_j}{\partial t^{k_0} \partial x_1^{k_1} \cdots \partial x_n^{k_n}}, \ldots \right), \tag{8}$$

$$i, j = 1, 2, \ldots, N; \; k_0 + k_1 + \cdots + k_n = k \leq n_j; \; k_0 < n_j.$$

$$\left(\frac{\partial^{k_0} u_i}{\partial t^{k_0}} \right)_{t=t_0} = \phi_i^{(k_0)}(x_1, \ldots, x_n). \tag{9}$$

All the functions are defined in the same region $G(x_1, \ldots, x_n)$. Cauchy's problem consists of finding a solution of system (8) with initial conditions (9). Now, Kovalevskaya's theorem is formulated as follows:

If all the functions F_i are analytic in a certain neighborhood of the point

$$\left(t^0, x_1^0, \ldots, x_n^0, \phi_1^{(0)}(x_1^0, \ldots, x_n^0), \ldots, \phi_N^{(0)}(x_1^0, \ldots, x_n^0), \ldots, \right.$$

$$\left. \frac{\partial^{k_1 + \cdots + k_n} \phi_j^{(k_0)}(x_1^0, \ldots, x_n^0)}{\partial x_1^{k_1} \cdots \partial x_n^{k_n}}, \ldots \right)$$

and all the functions $\phi_j^{(k_0)}(x_1, \ldots, x_n)$ are analytic in a neighborhood of the point $(x_1^0, \ldots, x_n^0)$, then Cauchy's problem has an analytic solution in a certain neighborhood of the point $(t^0, x_1^0, \ldots, x_n^0)$ and, moreover it is the unique solution in the class of analytic functions.

This result can certainly be derived from Kovalevskaya's work, but one should remember that the notion of the class of analytic functions, considered as a space in its own right, and the concepts of existence and uniqueness as understood in modern set theory, were new in Kovalevskaya's time. She does not express herself in this language, which, indeed is somewhat later than the publication date of her paper. To Kovalevskaya the important thing about her paper was the justification it gave for using the Cauchy problem as the definition of an analytic function.

The second remark, with which this chapter will come to a close, concerns one aspect of the nineteenth-century style of writing proofs which Kovalevskaya shared with her contemporaries, but which is definitely illogical. In the first section of the paper just discussed she needed to show that the function

$$\psi = \frac{1 - (1 - b)y - ax - \sqrt{(1 - (1 + b)y - ax)^2 - 4abx}}{2(1 - y)}$$

satisfies the partial differential equation and initial condition

$$\frac{\partial \psi}{\partial x} = \frac{a}{1 - \psi} \frac{\partial \psi}{\partial y}, \tag{10}$$

$$\psi(0, y) = \frac{by}{1 - y}. \tag{11}$$

Now this verification is something any calculus student can do. All one has to do is calculate the derivatives occurring in the equation, make the appropriate substitutions, and observe that the resulting equation is an identity in x and y. Kovalevskaya did not follow this procedure, however. Instead she wrote that equation (10) implies some dependence between the quantities ψ and $(1 - \psi)y + ax$ and that the initial condition (11) then proves that the connection between ψ, x, and y is given by the equation

$$(1 - \psi)y + ax = \frac{1 - \psi}{b + \psi} \psi,$$

from which the stated formula for ψ follows.

Since the solution is correct, it is perhaps unkind to point out that all this discussion of the manner in which the solution was obtained is unnecessary. It is important to note, however, that it is also not sufficient. The only real proof which she needed to give was the verification described above. She was not the only mathematician who wrote in this manner. For instance, Cauchy's verification of his solution to the quasilinear equation discussed in Section 2.3 is done exactly the same way. The logical gap in the argument will be of importance in the paper to be discussed in Chapter 6.

CHAPTER 3

Degenerate Abelian Integrals

3.1. Introduction

Kovalevskaya's work on Abelian integrals which reduce to elliptic integrals (1884) is the hardest of her papers to explain to a general audience since it assumes a detailed knowledge of the nineteenth-century work in this area. The subject of Abelian integrals in the nineteenth century was a vast corpus of results, many of which are not generally taught nowadays, even to special-ists in algebraic functions. This chapter therefore requires more preliminary explanation than the other chapters in this book, partly in order to introduce the reader properly to the mathematical concepts involved and partly in order to establish the historical context for Kovalevskaya's work. Appendix 2 contains a little analytic function theory for readers who have not had such a course. It is hoped that this appendix may clarify any mathematical ob-scurities in the exposition. Naturally this chapter is not a textbook on Abelian integrals. The reader who does not know the subject already should never-theless be able to appreciate in general terms the significance of the problem Kovalevskaya worked on. A really detailed exposition of the subject, even with mathematical details omitted, would have to be much longer than the present chapter. As a matter of fact, several such expositions were written around the beginning of the twentieth century, for instance Brill and Noether (1894) and Krazer and Wirtinger (1921). The latter occupies nearly 300 pages, proofs omitted. The former is some 450 pages long, and its con-tinuation by Emmy Noether in 1919 adds another 20 pages to its length. Expositions with complete proof by Weierstrass (*Werke* IV) and Baker (1897) occupy 600 and 700 pages, respectively. The most digestible introduction to one version of the theory is Bliss (1933). To see what the theory has become in the twentieth century, consult Forster (1981).

For the purposes of the present book Kovalevskaya's work in this area is important as an indication of the content of her education and as the work which reintroduced her to mathematical circles after a 6-year hiatus. More important still is the fact that the paper provides a clue to her approach to solving the Euler equations for a rotating rigid body. Indeed, if one were to choose one area to call Kovalevskaya's area of expertise, that area would be Abelian integrals. What unity there is in her rather diverse output is provided by this subject.

The first 11 sections of this chapter are devoted to establishing the mathematical background and historical context for Kovalevskaya's paper. The exposition is necessarily restricted to essentials and relies considerably on secondary sources, especially the papers of Brill and Noether and of Krazer

and Wirtinger mentioned above. Let us begin with some purely mathematical
background.

If $f(x, y)$ is a polynomial in two variables, the equation $f(x, y) = 0$
defines y implicitly as a multivalued function of x. A function $R(x, y)$ can then
be thought of as a multivalued function of x. It then makes sense to talk about
an integral $\int R(x, y) \, dx$. If $f(x, y)$ is irreducible and $R(x, y)$ is a rational function, such integrals are called Abelian integrals. The simplest Abelian integrals, those which lead to elementary and elliptic functions, are discussed in
Appendix 2. The rest of this chapter is written on the assumption that the
reader has some understanding of these simple integrals.

3.2.   Euler

According to Fricke (1913, p. 183), Euler was inspired by refereeing a paper
which Fagnano had sent to the Berlin Academy and began a study of the
integral

$$\int (1 - x^4)^{-1/2} \, dx.$$

In connection with this integral he studied the differential equation

$$m(1 - x^4)^{-1/2} \, dx = n(1 - y^4)^{-1/2} \, dy$$

and discovered (1752) the fact that

$$\int_0^z (1 - x^4)^{-1/2} \, dx + \int_0^w (1 - x^4)^{-1/2} \, dx = \int_0^R (1 - x^4)^{-1/2} \, dx,$$

where $R = (1 + z^2 w^2)^{-1}(z(1 - x^4)^{1/2} + w(1 - z^4)^{1/2})$. The general principle illustrated here is that the sum of two such integrals can be written as
a single integral whose upper limit is an algebraic function of the upper limits
of the other two. Note that this result would be expressed in the (later)
language of Abelian integrals introduced above by saying that

$$\int_0^{x_1} \frac{1}{y} \, dx + \int_0^{x_2} \frac{1}{y} \, dx = \int_0^R \frac{1}{y} \, dx,$$

where $f(x, y) = y^2 + x^4 - 1 = 0$ and $R = (x_1 y_2 + x_2 y_1)/(1 + x_1^2 x_2^2)$. The
existence of such algebraic addition theorems subsequently became one of the
important results about Abelian integrals.

3.3.   Legendre

The term "elliptic integral" was introduced by Legendre, who performed the
classifying and taxonomic work which, it can be seen in retrospect, shaped
the later development of the subject to a remarkable degree. Legendre's
results were published piecemeal in his *Exercises du Calcul Intégral*. In the
1820s, when he was already past 70, Legendre published a comprehensive

treatise on elliptic integrals summarizing his life's work, which he believed had brought the subject to the highest possible stage of development. As it turned out, however, the very act of setting all these ideas down in good order inspired the old man to make a new discovery, leading to still deeper levels of understanding.

The primary need in the early days of the subject was to reduce the number of different elliptic integrals which needed detailed study. Legendre did this by providing a classification of elliptic integrals at the very beginning of his treatise (1825, Traité, I; pp. 14–18). He showed that any integral of the form $\int (P/R) \, dx$, where P is a rational function and R the square root of a fourth-degree polynomial, could be reduced to an integral of the form

$$H = \int \frac{A + B \sin^2\phi}{1 + n \sin^2\phi} \frac{1}{\Delta} \, d\phi, \qquad \Delta = (1 - c^2 \sin^2\phi)^{1/2},$$

which he called an elliptic function or elliptic transcendant.

Having made this reduction, he then showed that any such integral can be written as a sum of elliptic functions of one of the following three types:

(1) $E = \int \Delta \, d\phi$ (obtained by taking $n = 0$, $A = 1$, $B = -c^2$);
(2) $\gamma = \Delta \tan \phi - E + b^2 \int (1/\Delta) \, d\phi$ (taking $A = b^2$, $B = -b^2 c^2$, $n = -1$);
(3) $\Pi = \int [1/(1 + n \sin^2 \phi) \Delta] \, d\phi$ (taking $A = 1$, $b = 0$).

Legendre immediately pointed out that it makes more sense to take as fundamental the function

$$F = \int \frac{1}{\Delta} \, d\phi$$

since, as he said, one can easily determine a function "in rational relation to a given function." That is, one can easily solve the equation $F(w) = rF(z)$ algebraically for w in terms of a given rational number r and a given value of z. For these reasons Legendre called F, E, and Π elliptic functions of first, second, and third kinds, respectively.

Legendre developed fully the properties of these three kinds of integrals and came extremely close to a systematic study of the inverse functions (which are the functions nowadays given the name elliptic functions). In fact he discovered the addition theorem for the inverse functions in disguised form. He showed (Traité I, p. 22) that if $F(\phi) + F(\psi) = F(\mu)$, then

$$\sin \mu = \frac{\sin(\phi)\cos(\psi)\,\Delta(\psi) + \cos(\phi)\sin(\psi)\,\Delta(\phi)}{1 - c^2 \sin^2\phi \, \sin^2\psi}.$$

Legendre was led to this result by studying the differential equation of Euler which he wrote in the form

$$\frac{d\phi}{\Delta\phi} + \frac{d\psi}{\Delta\psi} = 0.$$

Legendre also studied the complete elliptic integrals

$$F'(c) = \int_0^{\pi/2} \frac{d\phi}{\Delta(c, \phi)}, \qquad E'(c) = \int_0^{\pi/2} \Delta(c, \phi)\, d\phi$$

and discovered the relation now known as Legendre's relation (Traité I, p. 61)

$$F'(c)E'((1 - c^2)^{1/2}) + F'((1 - c^2)^{1/2})E'(c) - F'(c)F'((1 - c^2)^{1/2}) = \tfrac{1}{2}\pi.$$

Nowadays we write $F'(c) = K$, $F'((1 - c^2)^{1/2}) = K'$, $E'(c) = E$, $E'((1 - c^2)^{1/2}) = E'$, so that this relation is more familiarly known as

$$KE' + EK' - KK' = \tfrac{1}{2}\pi.$$

This relation later formed the basis for beautiful investigations of Weierstrass.

To reduce still further the number of different integrals one needs to evaluate, Legendre's treatise exhibits connections between the various elliptic functions corresponding to different values of the parameter c. For instance, Legendre showed (Traité I, p. 79) that

$$F(c', \phi') = \frac{1 + c}{2} F(c, \phi)$$

provided $c' = 2c^{1/2}/(1 + c)$ and $\sin(2\phi' - \phi) = c \sin \phi$. The theory of transformations thus generated resulted in considerable saving in labor when calculating tables of elliptic functions. A study of the integral for one value of c and all values of ϕ brought with it information about a whole class of such integrals, namely, those with parameters obtained from c by repeated application of the transformation $c \to c'$. The existence of such transformations of course implies something about the structure of elliptic functions. The elaboration of that structure was one of the principal achievements of Legendre's successors.

Legendre called the class of elliptic functions obtainable from a single function by repeated transformations a scale (échelle). In early 1825, as his treatise was nearing completion, he made the discovery of a new échelle leading to the transformation (Traité I, p. 224)

$$F(\alpha, \omega) = mF(c, \phi),$$

where moduli α and c amplitudes ω and ϕ are related by

$$1 - \alpha^2\sin^2\omega = (1 - c^2\sin^2\phi)\left(\frac{1 - (k/m)\sin^2\phi}{1 + k\sin^2\phi}\right)^2,$$

$$k = \tfrac{1}{4}(m - 1)(m + 3),$$

for $1 < m < 3$.

This ability of the function F to reproduce itself under two different kinds of transformation so impressed Legendre that he wrote a supplement to his treatise to explore it more fully. Indeed he even suggested applying Bernoulli's epithet on the logarithmic spiral (Eadem mutata resurgit) to the function F.

We may summarize the points of the first volume of Legendre's treatise which are important for present purposes as the classification of the three kinds of elliptic functions (a trichotomy which remains valid for arbitrary Abelian integrals), the Legendre relation for the complete elliptic integrals, and the theory of transformations. The treatise, of course, contains much that is valuable besides these things. Legendre developed the properties of elliptic functions in great detail and gave many applications. He even considered the case where the parameter n in the integral of the third kind assumes imaginary values, but he did not investigate his elliptic functions as complex integrals. Legendre, as mentioned, was very old, and Cauchy's theory of complex integration was very young, in 1825. Although Legendre had reviewed Cauchy's 1814 paper which introduces complex integrals (cf. Grattan-Guinness 1982, 9.2.5), the theory presented in that paper was very rudimentary. For example, the limits of integration were real, the poles were simple poles, etc.

Among the applications, the study of the equations of motion of a rigid body rotating about a fixed point occupies a prominent place. Legendre studied many cases of this motion, paying special attention to the two cases now known as the Euler case and the Lagrange case (Traité I, pp. 366–410). The fact that elliptic integrals arise in these special cases suggests that the general case involves even more complicated integrals. Thus one can already see some physical motivation beyond the mathematical motivation for extending Legendre's investigations to more complicated integrals.

Legendre's second volume is devoted to the construction of tables of the elliptic functions and their use in certain applications, chiefly to the Eulerian integrals (what are more commonly known nowadays as beta and gamma functions.)

Despite the date of 1825 which appears on the title page of Legendre's volume I, he tells us in the *avertissement* to the third volume that the first volume did not appear until January 1827. Legendre was therefore astonished to find that his discovery of the second *échelle* had been duplicated by Jacobi in letters which Legendre says were sent to Schumacher on 13 June and 2 August 1827 and printed in Vol. 6 of the *Astronomische Nachrichten*. Indeed Jacobi had discovered other transformations as well. Meanwhile Abel also had begun to study elliptic integrals, and, using a crude type of complex integral (actually proceeding formally with imaginary values substituted in a real integral), had discovered some of the properties which these functions have as functions of a complex variable. Legendre gave the highest praise to these two young mathematicians and devoted his third volume to an exposition of some of their results and a reevaluation of the subject in light of these results. Although his reputation has been somewhat eclipsed by the spectacular work of Abel and Jacobi, one can nevertheless see his influence in the problems which later mathematicians formulated as fundamental to the subject of Abelian integrals. The subject developed very much along lines laid down by Legendre for the special case of elliptic integrals; and if that direction of development is forced by the nature of the subject, still Legendre must be given credit for perceiving that structure already in the case of elliptic integrals.

3.4. Abel

The man for whom Abelian integrals are named devoted much attention to both elliptic integrals and their generalizations. In his papers on elliptic integrals (1827, 1828) he considered the integral

$$\alpha = \int_0^x (1 - c^2 x^2)^{-1/2} (1 + e^2 x^2)^{-1/2} \, dx.$$

This particular form of the elliptic integral was evidently chosen because the integral is simply related to the one which results when x is replaced by ix where $i = \sqrt{(-1)}$. Abel took the important step of regarding x as a function $\phi(\alpha)$ of α in this equation. If β represents the integral which results when the plus and minus signs in the integrand are reversed, one then has $ix = \phi(\beta i)$. Setting $f(\alpha) = (1 - c^2 \phi^2(\alpha))^{1/2}$ and $F(\alpha) = (1 + e^2 \phi^2(\alpha))^{1/2}$, Abel obtained Legendre's addition formula in the form

$$\phi(\alpha + \beta) = \frac{\phi(\alpha) f(\beta) F(\beta) + \phi(\beta) f(\alpha) F(\alpha)}{1 + e^2 c^2 \phi^2(\alpha) \phi^2(\beta)}$$

(see Abel 1827, p. 105, or Oeuvres I, p. 268).

   With the value of the function established for purely imaginary values of the argument and this addition formula, Abel was equipped to study the function as a function of a complex variable. One of his first discoveries was the general solution of the equation $\phi(x) = \phi(\alpha)$, namely,

$$x = (-1)^{m+n} \alpha + m\omega + n\tilde{\omega}i,$$

(1827, p. 114, or Oeuvres I, p. 278) where m and n are arbitrary integers and

$$\omega = 2 \int_0^{c^{-1}} ((1 - c^2 x^2)(1 + e^2 x^2))^{-1/2} \, dx,$$

$$\tilde{\omega} = 2 \int_0^{e^{-1}} ((1 - e^2 x^2)(1 + c^2 x^2))^{-1/2} \, dx.$$

This discovery established the double periodicity of elliptic functions and showed the importance of complex variables for handling them.

   Abel had found an important tool for working with integrals of algebraic functions, and he was not slow to exploit it in more general contexts than that of elliptic functions. His thought is neatly summarized in his own words (Oeuvres I, p. 444):

   If ψx denotes the most general elliptic function, that is if

$$\psi x = \int \frac{r \, dx}{\sqrt{R}}$$

where r is an arbitrary rational function of x and R a polynomial in the same variable of degree not larger than 4, this function has, as is known, the very remarkable property that the sum of an arbitrary number of these functions can be expressed by a single function of the same form upon the addition of a certain algebraic and logarithmic expression.

It seems that in the theory of transcendental functions mathematicians have limited themselves to functions of this form. Yet there exists for a very extensive class of other functions a property analogous to that of the elliptic functions.

I have in mind those functions which can be regarded as *integrals of arbitrary algebraic differentials*. While one cannot express the sum of an arbitrary number of given functions by a single function of the same type, as is the case with elliptic functions, one can at least always express such a sum by the sum of a fixed number of other functions of the same nature as the original ones upon adding a certain algebraic and logarithmic expression. We shall demonstrate this property in a future issue of this journal. For the moment I shall consider a particular case which includes the elliptic functions, namely the functions contained in the formula

$$\psi x = \int \frac{r\,dx}{\sqrt{R}},$$

R being an arbitrary polynomial and r a rational function.

Abel's particular case is the one now called hyperelliptic, i.e., the case where x and y are related by $y^2 = p(x)$, p being a polynomial. As the quotation shows, Abel had results which applied not only to this special case, but to an arbitrary algebraic differential. His theorem reads as follows (Oeuvres I, p. 454):

Let $\psi x = \int r\,dx/\sqrt{\phi x}$, where r is an arbitrary rational function and ϕx a polynomial of degree $2\nu - 1$ or 2ν; and let $x_1, x_2, \ldots, x_{\mu_1}, x'_1, x'_2, \ldots, x'_{\mu_2}$ be given variables. Whatever the number $\mu_1 + \mu_2$ of variables, one can always find by means of an algebraic equation $\nu - 1$ quantities $y_1, y_2, \ldots, y_{\nu-1}$ such that

$$\psi x_1 + \psi x_2 + \cdots + \psi x_{\mu_1} - \psi x'_1 - \psi x'_2 - \cdots - \psi x'_{\mu_2}$$

$$= v + \epsilon_1 y_1 + \epsilon_2 y_2 + \cdots + \epsilon_{\nu-1} y_{\nu-1}$$

v being algebraic and logarithmic and $\epsilon_1, \epsilon_2, \ldots, \epsilon_{\nu-1}$ equal to $+1$ or -1.

As a first approximation we may take this theorem to say that, although the indefinite integral of an algebraic function is not usually algebraic itself, the amount of "transcendence" which one can generate using a single integrand and different upper limits of integration is limited. Except for the function v any sum "collapses" to a sum of $\nu - 1$ integrals with upper limits algebraically dependent on those of the original sum. In its general form this theorem is known as Abel's theorem. The minimum number of integrals to which a sum of integrals can be algebraically reduced plays a very important role and arises in many different disguises, some of which will be seen below. Abel remarked that the functions y_j are unchanged whatever the form of the rational function r and that v remains the same when a polynomial of degree $\nu - 2$ is added to r.

3.5. Cauchy

A deeper insight into the periodicity of the inverses of Abelian integrals was achieved through the work of Cauchy on complex integration. According to Brill and Noether (1894, pp. 155–159), some of Cauchy's work in this area was anticipated by Gauss in a letter sent to Bessel on 12 January 1812 (*Briefwechsel zwischen Gauss und Bessel,* Leipzig 1880 = *Werke* III, p. 156). In this letter Gauss gave a definition along the lines of "Riemann sums" for the integral $\int \phi(x) \, dx$ when x is a complex variable. He noticed that the integral $\int 1/x \, dx$ increases by $2\pi i$ each time the path of integration encircles the point 0. He thereby discovered the multivaluedness of the logarithm which is reflected in the periodicity of the inverse function (the exponential).

Cauchy's first work on "complex integration" appeared in 1814. According to Grattan-Guinness (1982, 9.2.4) Cauchy's integrals did not differ essentially from real integrals in this first paper. Cauchy did, however, discover the complex plane and a limited "residue" theorem for simple poles.

A further significant step was taken by Cauchy in 1846. In that year, say Brill and Noether (1894, pp. 165, 173ff.), Cauchy integrated *multivalued* functions around a closed curve, thereby obtaining what he called "indices de périodicité," which unlike the periodicity exhibited by single-valued functions such as the integral considered by Gauss, could not be represented as "residues," i.e., as $\lim z \rightarrow a \, 2\pi i (z - a) f(z)$, but had to be expressed as integrals along lines joining the points of discontinuity. These periods, later called "moduli of periodicity," play an important role in the systematic study of Abelian integrals.

3.6. Jacobi

It was noted above that Jacobi studied elliptic functions at the beginning of his career. It is to him that we owe much of the notation now used in the subject of what are called Jacobian elliptic functions (as opposed to the Weierstrassian elliptic functions). In fact (cf. Fricke 1913, pp. 202–209), given the equation

$$\Xi = \int_0^\phi (1 - k^2 \sin^2 \phi)^{-1/2} \, d\phi$$

Jacobi wrote $\phi = am(\Xi, k)$ and introduced the function $x = \sin \phi = \sin am(\Xi, k)$, as well as $\cos am(\Xi, k) = (1 - x^2)^{1/2}$ and $\Delta \, am(\Xi, k) = (1 - k^2 x^2)^{1/2}$. According to Weierstrass (*Werke* I, p. 120) these notations were changed to sn, cn, dn by Gudermann. Like Abel, Jacobi attempted to extend his study to irrationalities more complicated than the elliptic case.

Jacobi noticed (1832b, p. 394) that the addition formula of Euler implies an algebraic addition formula for elliptic functions, i.e., there exists a nonzero polynomial $q(x, y, z)$ such that $q(f(x), f(y), f(x+y)) = 0$ for all x and y when f is an elliptic function. Jacobi asked what Abel's generalization of this addition theorem implies about the inverse function of an Abelian integral. He had found (1828, p. 310) that a single hyperelliptic integral cannot "really" be inverted. In fact he noted that if y is the square root of a polynomial of degree 5 or 6, then the integral

$$w = \int_a^z \frac{\alpha + \beta x}{y}\, dx$$

has more than two independent periods. (It will be recalled that Abel found essentially two independent periods 2ω and $2\bar{\omega}i$ for his elliptic integrals.) Jacobi showed that by altering the path from a to z he could make the integral for w approach as close as desired to any complex number whatsoever. He considered that this argument proved the impossibility of functions having more than two independent periods. This argument is of course valid (and indeed still presented today in courses on complex analysis), but it applies only to single-valued functions. As Weierstrass pointed out in a letter to Kovalevskaya of 24 March 1885 (Kochina 1973, p. 120), an infinitely-many-valued function can have more than two independent periods.

The fact that the inverse functions of algebraic functions are simpler than the integrals themselves was so well established by the work of Abel on elliptic functions that it was almost inevitable that this route would be followed in the study of more complicated irrationalities. One possible route, as just seen, would have been the use of infinitely-many-valued functions. Jacobi apparently could not have used this route. Fortunately he found another equally fruitful method. If one attempts to classify hyperelliptic integrals as Legendre classified elliptic integrals, one notices that the elliptic integrals of the first kind have the property of remaining bounded for all ranges of integration. If one is dealing with hyperelliptic integrals involving the square root of a polynomial $\phi(x)$ of degree $2\nu - 1$ or 2ν, the only such integrals are of the form

$$\int \frac{p(x)}{\sqrt{\phi(x)}}\, dx,$$

where $p(x)$ is a polynomial of degree, at most $\nu - 2$. There are thus only $\nu - 1$ independent integrals of the first kind when the irrationality is the square root of a polynomial of degree $2\nu - 1$ or 2ν, namely,

$$\Phi_k(z) = \int_0^z \frac{x^k}{y}\, dx, \qquad k = 0, 1, \ldots, \nu - 2,$$

where $y^2 = \phi(x)$ and ϕ is of degree $2\nu - 1$ or 2ν. Notice that the number $\nu - 1$, which occurred in Abel's theorem as the maximal number of integrals

needed to express a sum of such integrals algebraically, now appears again
as the number of independent integrals of the first kind. Jacobi pursued the
inverse functions by posing the problem of solving the equations

$$u_0 = \Phi_0(x_0) + \Phi_0(x_1) + \cdots + \Phi_0(x_{\nu-2}),$$
$$u_1 = \Phi_1(x_0) + \Phi_1(x_1) + \cdots + \Phi_1(x_{\nu-2}),$$
$$\vdots$$
$$u_{\nu-2} = \Phi_{\nu-2}(x_0) + \cdots + \Phi_{\nu-2}(x_{x-2})$$

for $x_0, \ldots, x_{\nu-2}$ in terms of $u_0, \ldots, u_{\nu-2}$. The problem posed here (1832b)
for hyperelliptic integrals easily extends to general Abelian integrals, and
became the subject of intensive research for the next quarter century. It is
known as the Jacobi inversion problem. Although he did not solve the problem, Jacobi did contribute to the solution in two different ways. First he
proved that the inverse functions possess an algebraic addition theorem
(1832b). Second, as already noted, he studied the inverse functions for the
case of elliptic integrals, where $\nu = 2$. In so doing he discovered a way to
represent these functions as quotients of standard "well-behaved" functions
which he called theta functions. Such a representation is needed since a
doubly-periodic function (not a constant) must have singularities. A typical
such representation is the following (Jacobi, *Fundamenta*, Art. 35 ff.; cited
in Fricke 1913, p. 213):

$$\sin am\left(\frac{2Kx}{\pi}\right) = k^{-1/2} \frac{2\sqrt[4]{q}\,(\sin x) \prod_{n=1}^{\infty}(1 - 2q^{2n}\cos 2x + q^{4n})}{\prod_{n=1}^{\infty}(1 - 2q^{2n-1}\cos 2x + q^{4n-2})}, \quad (1)$$

where K and K' have the meanings given in Section 3.3 and $q = e^{-\pi K'/K}$. The
numerator and denominator of the fraction on the right-hand side of (1) are
not doubly periodic. They each have period 2π, as does the function on the
left-hand side of (1). When x is increased by a multiple of $i\pi K'/K$, the
left-hand side of (1) remains unchanged. The numerator and denominator of
the right-hand side are each multiplied by the same constant, which therefore
cancels out, leaving the right-hand side unchanged as well. The functions in
the numerator and denominator, the theta functions, thus provide the solution
to the problem posed by the paucity of doubly-periodic functions without
singularities. They are in an intuitive sense "sesqui-periodic," in that they
have one authentic period, and one "pseudo-period," under which they are
reasonably behaved. Suitably generalized, they provide as good a representation as can be expected for the inverse functions demanded by the Jacobi
inversion problem.

3.7. Göpel and Rosenhain

The first solution of a special case of the Jacobi inversion problem seems to have been achieved by Göpel for the case $\nu = 3$ (1847), for which there are two independent variables. Göpel's work was extended by Rosenhain (1850). According to Krazer and Wirtinger (1921, p. 619) these two authors discovered a suitable generalization of theta functions for two variables. Sixteen functions of this form turned out to be of particular use, since 10 of them were even functions and the other 6 odd. If one of these functions is singled out and used as denominator (there is some freedom in the choice of the function so used) the inverse functions for this case can be expressed in terms of the 15 quotients thereby produced. Thus Göpel and Rosenhain had shown that if

$$u = \int_{x_0}^{x} \frac{ds}{\sqrt{\phi(s)}} + \int_{y_0}^{y} \frac{ds}{\sqrt{\phi(s)}},$$

$$v = \int_{x_0}^{x} \frac{s\,ds}{\sqrt{\phi(s)}} + \int_{y_0}^{y} \frac{s\,ds}{\sqrt{\phi(s)}},$$

(2)

where ϕ is a polynomial of degree 5 or 6, then any symmetric function of x and y is a single-valued function of u and v. [Obviously, since u and v are symmetric functions of x and y if $x_0 = y_0$, any function of u and v must be a symmetric function of x and y. The importance of the result of Göpel and Rosenhain is the converse. In modern language it allows a mapping of (x, y) to be "lifted" to a mapping of (u, v).] The explicit connection with the Jacobi inversion problem was made by Rosenhain in the paper cited, where he showed that if (u, v) are taken as arguments in the 15 quotients, then all 15 can be written using the 2 variables x and y connected with u and v by (2).

3.8. Weierstrass' Early Work

Integrals involving the square root of a polynomial of arbitrary degree were studied by Weierstrass in a manuscript dated 17 July 1849 and published as an addendum to the 1848–1849 annual report of the Braunsberg Gymnasium (*Werke* I, pp. 111–131). As Weierstrass said in the introduction to this paper, he had been trying for some time to solve the Jacobi inversion problem, that is, to find an explicit representation of the inverse functions of the integrals of first kind. His method was developed in close analogy with the theory of elliptic functions.

Weierstrass' point of departure was the Legendre relation

$$KE' + EK' - KK' = \tfrac{1}{2}\pi,$$

for which he found an analog in the equation

$$\int_{a_\mu}^{a_{\mu+1}} \int_{a_\nu}^{a_{\nu+1}} \frac{F(x, y)\,dx\,dy}{\sqrt{R(x)}\,\sqrt{R(y)}} = \begin{cases} 0 & \text{if } \nu > \mu + 1 \\ \dfrac{\pi}{2i} & \text{if } \nu = \mu + 1. \end{cases}$$

Here $R(x) = (x - a_1)(x - a_2) \cdots (x - a_{2n})(x - a_{2n+1})$, with $a_1 < a_2 < \cdots < a_{2n+1}$ and

$$F(x, y) = \frac{1}{2} \frac{R(x) - R(y)}{(x - y)^2} - \frac{1}{4} \frac{R'(x) - R'(y)}{x - y}.$$

He posed the Jacobi inversion problem in a normalized form requiring the inversion of the system

$$u_j = \int_{a_1}^{x_1} \frac{F_j(x)}{\sqrt{R(x)}} \, dx + \int_{a_3}^{x_2} \frac{F_j(x)}{\sqrt{R(x)}} \, dx + \cdots + \int_{a_{2n-1}}^{x_n} \frac{F_j(x)}{\sqrt{R(x)}} \, dx,$$

$j = 1, 2, \ldots, n$. Here $F_j(x)$ is a constant multiple of

$$\frac{(x - a_1)(x - a_3) \cdots (x - a_{2n+1})}{x - a_{2j-1}}.$$

This formulation of the problem enabled Weierstrass to introduce n single-valued functions $p_1, \ldots, p_n$ analogous to sin $am\ u$; in fact p_j was a constant multiple of $\sqrt{(a_{2j-1} - x_1)(a_{2j-1} - x_2) \cdots (a_{2j-1} - x_n)}$ and was denoted $sn(u_1, \ldots, u_n)_j$. A similar expression with a_{2j-1} replaced by a_{2j} was denoted $cn(u_1, \ldots, u_n)_j$. Finally, replacing a_{2j-1} by a_{2n+1} gave an analog for $\Delta\ am\ u$, denoted $dn(u_1, \ldots, u_n)$. When $n = 1$, the functions thus introduced actually do reduce to the Jacobian elliptic functions. The analog of the complete elliptic integrals K and K' of first kind was provided by the $2n^2$ numbers

$$K_{rs} = \int_{a_{2s-1}}^{a_{2s}} \frac{F_r(x)}{\sqrt{R(x)}} \, dx,$$

$$K'_{rs} = \sum_{t=s}^{n} i \int_{a_{2t}}^{a_{2t+1}} \frac{F_r(x)}{\sqrt{R(x)}} \, dx.$$

Weierstrass showed that the numbers

$$\omega_a = m_1 K_{a1} + \cdots + m_n K_{an} + (r_1 K'_{a1} + \cdots + r_n K'_{an})i$$

were "semi-periods" (not Weierstrass' name for them) in the sense that

$$sn(u_1 + 2\omega_1, u_2 + 2\omega_2, \ldots, u_n + 2\omega_n)_j$$

$$= (-1)^{m_j + r_1 + \cdots + r_{j-1}} sn(u_1, u_2, \ldots, u_n)_j.$$

After finding $2n^2$ numbers J_{rs} and J'_{rs} which were similarly analogous to the complete elliptic integrals of second kind E and E', Weierstrass used (3) to deduce the analog of Legendre's relation, namely,

$$\sum_a (K_{ab} J_{ac} - J_{ab} K_{ac}) = 0,$$

$$\sum_a (K'_{ab} J'_{ac} - J'_{ab} K'_{ac}) = 0,$$

$$\sum_a (K_{ab}J'_{ac} - J_{ab}K'_{ac}) = 0 \qquad \text{if } b \neq c,$$

$$\sum_a (K_{ab}J'_{ab} - J_{ab}K'_{ab}) = \frac{\pi}{2}.$$

Using these relations Weierstrass constructed an auxiliary function (*Hülfsfunction* in the German spelling of the time) similar to Jacobi's theta functions. He showed that if t_a is chosen as a suitable linear combination of $u_1, \ldots, u_n$, then the functions $\text{sn}(u_1, \ldots, u_n)_j$, $\text{cn}(u_1, \ldots, u_n)_j$, and $\text{dn}(u_1, \ldots, u_n)$ can all be expressed as quotients of two such functions, which he denoted $\text{Hl}(t_1, \ldots, t_n)$. He promised to give the details of his arguments in a subsequent paper.

The subsequent paper was one of Weierstrass' most famous and led to his being called to Berlin. It appeared in Bd. 47 of Crelle's *Journal für die reine und angewandte Mathematik* (cf. *Werke* I, pp. 133–152). Its date of submission was given as 11 September 1853. A note from the editor stated that the results therein contained had already appeared in a manuscript of the author dated 17 July 1849. In this paper Weierstrass promised much more than he had done in the 1849 paper just discussed. He stated his intention to publish a series of papers on Abelian transcendents, beginning with the present one, in which he would develop the properties of periodic functions of several variables in a way quite different from that of Göpel and Rosenhain and which would suggest the possibility of handling higher transcendents as well, i.e., irrationalities more complicated than the square root of a polynomial.

This 1854 paper retraced the results of the 1849 paper using more systematic notation and developing the theta functions in more detail. In it Weierstrass used the notation $R(x) = (x - a_0)(x - a_1) \cdots (x - a_{2n})$ and denoted the functions $\text{sn}(u_1, \ldots, u_n)_1, \ldots, \text{sn}(u_1, \ldots, u_n)_n$, $\text{cn}(u_1, \ldots, u_n)_1, \ldots, \text{cn}(u_1, \ldots, u_n)_n, \text{dn}(u_1, \ldots, u_n)$ by $\text{al}(u_1, \ldots, u_n)_0, \ldots, \text{al}(u_1, \ldots, u_n)_{n-1}, \text{al}(u_1, \ldots, u_n)_n, \ldots, \text{al}(u_1, \ldots, u_n)_{2n-1}, \text{al}(u_1, \ldots, u_n)_{2n}$, respectively. The letters al of course were chosen to stand for Abel. Weierstrass called these functions Abelian functions. He found representations for them as quotients of theta functions which he denoted $\text{Jc}(v_1, \ldots, v_n)$, the letters of course in honor of Jacobi.

The series of papers Weierstrass had planned turned out to be rather short. In fact only one more paper appeared, in Bd. 52 of Crelle's *Journal* (1856b or *Werke* I, pp. 297–355). This paper was a rather comprehensive treatise on the Jacobi inversion problem for hyperelliptic integrals, with notation once again modified. This time setting $R(x) = A_0(x - a_1)(x - a_2) \cdots (x - a_{2p+1})$, $P(x) = (x - a_1)(x - a_2) \cdots (x - a_p)$, $Q(x) = A_0(x - a_{p+1}) \cdots (x - a_{2p+1})$, Weierstrass posed the Jacobi inversion problem as a system of differential equations

$$du_j = \frac{1}{2} \frac{P(x_1)}{x_1 - a_j} \frac{1}{\sqrt{R(x_1)}} dx_1 + \cdots + \frac{1}{2} \frac{P(x_p)}{x_p - a_j} \frac{1}{\sqrt{R(x_p)}} dx_p,$$

$j = 1, 2, \ldots, p$, with initial conditions $x_j(0, 0, \ldots, 0) = a_j$. He showed
that the solutions were the roots of the equation

$$\sum_{j=1}^{p} \left\{ \frac{Q(a_j)}{P'(a_j)} \frac{al^2(u_1, \ldots, u_p)_j}{x - a_j} \right\} = 1.$$

He had thus shown how to express the functions of the Jacobi inversion
problem algebraically in terms of standard functions. Shortly after this paper
appeared, Weierstrass' research took a new direction, causing him to withdraw his next paper on this subject after he had submitted it for publication.
The cause of this change of direction will be seen below. Before we turn to
that subject, however, there is one more development of importance which
occurred shortly after the appearance of the paper just discussed. We must
therefore take up that topic first.

3.9. Hermite

Hermite conducted an investigation of the general Jacobi inversion problem;
more precisely, the problem of finding the inverse functions for a complete
system of Abelian integrals of first kind. Indeed, he attempted to invert more
general systems of integrals, where the irrationality is not necessarily the
same in all the integrals. These results were communicated in a letter to
Liouville and published in the *Comptes Rendus* volume XVIII (1844 or
Oeuvres I, pp. 49–63). He found, as he said, "the greatest difficulties" in this
investigation and carried it out in detail only for elliptic integrals.

After the appearance of Weierstrass' paper in 1854 Hermite began to see
the light. He performed a reprise of the work of Göpel and Rosenhain using
some of Weierstrass ideas and an approach which he originated. In this work
(1855 or *Oeuvres* I, pp. 444–478) he considered along with system (2) of
Göpel and Rosenhain a second system

$$u = \int_{x_0}^{x} \frac{(\alpha + \beta s)\, ds}{\sqrt{\psi(s)}} + \int_{y_0}^{y} \frac{(\alpha + \beta s)\, ds}{\sqrt{\psi(s)}},$$

$$v = \int_{x_0}^{x} \frac{(\gamma + \delta s)\, ds}{\sqrt{\psi(s)}} + \int_{y_0}^{y} \frac{(\gamma + \delta s)\, ds}{\sqrt{\psi(s)}}.$$

(3)

Denoting the 15 quotients found by Göpel and Rosenhain for system (2)
by $f_1(u, v), \ldots, f_{15}(u, v)$ and the 15 for system (3) by $F_1(u, v), \ldots,$
$F_{15}(u, v)$, Hermite asked for which integrands in system (3) the functions F_j
are rational functions of the f_j. The formulation of this problem can be seen
as an attempt to extend Legendre's transformation of moduli to hyperelliptic
functions, though of course the problem had been modified since Legendre's
day by the emergence of theta functions as a primary tool. If F_j is a rational
function of the f_j, then any common period of the f_j must be a period of F_j.
When he expressed each common period of the f_j as a sum of integral
multiples of the four periods of F_j, Hermite obtained an arithmetic theory of

transformation of theta functions. Although the details require too much calculation to be given here, it will be useful to set down a few of the more prominent equations as a basis for discussing the further work of Riemann and Weierstrass.

Hermite began by considering typical functions $F(u, v)$ and $f(u, v)$. These functions are quadruply periodic. Let the periods of F be (Ω_j, γ_j), $j = 0, 1, 2, 3$; and let the periods of $f(u, v)$ be (ω_j, ν_j), $j = 0, 1, 2, 3$. Since each period of $f(u, v)$ must be a period of $F(u, v)$ there must be integers $a_0, \ldots, a_3, b_0, \ldots, b_3, c_0, \ldots, c_3, d_0, \ldots, d_3$ such that

$$\omega_0 = a_0\Omega_0 + \cdots + a_3\Omega_3,$$

$$\nu_0 = a_0\gamma_0 + \cdots + a_3\gamma_3,$$

$$\vdots$$

$$\omega_3 = d_0\Omega_0 + \cdots + d_3\Omega_3,$$

$$\nu_3 = d_0\gamma_0 + \cdots + d_3\gamma_3.$$

From this fact and other known relations among the periods, it followed that the determinant of the system of equations for the $\omega_j's$ must be k^2, where k is the integer $a_0d_3 - d_0a_3 + b_0c_3 - c_0b_3$. The function $f(x, y)$ given by

$$f(x, y) = F(\Omega_0 x + \Omega_1 y, \gamma_0 x + \gamma_1 y)$$

then has periods $(1, 0)$, $(0, 1)$, (H, G'), and (G, H), where

$$G = \frac{\Omega_3\gamma_1 - \Omega_1\gamma_3}{\Omega_0\gamma_1 - \Omega_1\gamma_0}; \qquad H = \frac{\Omega_2\gamma_1 - \Omega_1\gamma_2}{\Omega_1\gamma_0 - \Omega_0\gamma_1}; \qquad G' = \frac{\Omega_0\gamma_2 - \Omega_2\gamma_0}{\Omega_0\gamma_1 - \Omega_1\gamma_0}.$$

Because the periods of f are simpler than those of the original functions, it is possible to express f as a quotient of theta functions. To this end Hermite set $\Phi(x, y) = Gx^2 + 2Hxy + G'y^2$, then defined 16 theta functions (the same ones found by Rosenhain) in a single formula

$$\Theta(x, y) = \sum (-1)^{mq+np} e^{i\pi[(2m+\mu)x+(2n+\nu)y+\Phi(2m+\mu,\, 2n+\nu)/4]},$$

where the summation extends over all integer values of m and n. The four integers p, q, μ, ν are each fixed at one of the values 0 or 1, thus giving 16 possible functions. It is easy to see that the 16 functions are closely related. In fact if $\Theta_0(x, y)$ is the one obtained by setting all four integers equal to 0, then any other function of this type satisfies

$$\Theta(x, y) = e^{i\pi[\mu x + \nu y + \Phi(\mu,\, \nu)/4]}\Theta_0\left(x + \frac{\mu G + \nu H + q}{2}, y + \frac{\mu H + \nu G' + p}{2}\right).$$

Hermite showed that a modification of Θ, namely,

$$\Pi(x, y) = \Theta(z_0 + Gz_3 + Hz_2, z_1 + Hz_3 + G'z_2)e^{i\pi(z_0z_3+z_1z_2 + \Phi(z_3,\, z_2))}$$

$(z_j = a_j x + b_j y)$ could be expressed as a homogeneous polynomial of degree k in four other theta functions Θ_0, Θ_1, Θ_2, Θ_3 whose moduli g, h, g' were rational functions of degree 2 in G, H, G'. Indeed two of the functions could be arbitrarily chosen, and the four functions satisfy a homogeneous equation of degree 4. The integer k came to be called the degree of the period transformation, since it is the degree of the polynomial which expresses Π in terms of the Θ_j.

Proceeding along these lines, Hermite answered in principle the question when the periodic functions which are the inverses of one hyperelliptic integral can be expressed rationally in terms of the inverses of another. The path he followed, as outlined above, led to relations among the quadruply-periodic quotients of the theta functions. But, as he said, such functions are the most general periodic functions generated by hyperelliptic integrals containing the square root of a fifth- or sixth-degree polynomial, and therefore it suffices to know when one theta function transforms rationally into another.

Hermite went on to show that the Weierstrassian functions al_j, $j = 0$, 1, 2, 3, 4 which are defined for this case (polynomial of degree five or six) could be expressed in terms of his 16 theta functions, as follows:

$$al_0 = \frac{\Theta_{1000}}{\Theta_{0000}}, \quad al_1 = \frac{\Theta_{1001}}{\Theta_{0000}}, \quad al_2 = \frac{\Theta_{0101}}{\Theta_{0000}}, \quad al_3 = \frac{\Theta_{0111}}{\Theta_{0000}}, \quad al_4 = \frac{\Theta_{0011}}{\Theta_{0000}},$$

where Θ_{abcd} represents Hermite's theta function with $p = a$, $q = b$, $\mu = c$, $\nu = d$.

3.10. Riemann

Weierstrass seemed to be closing in on the general solution of the Jacobi inversion problem in the mid-1850s. His work was being followed with great interest and respect by Riemann, who was at the time Privatdozent in Göttingen. Riemann himself had been interested in the problem for several years and had been applying his own version of complex function theory in an attempt to solve it. The results of this investigation, using the Riemann surface to study algebraic singularities (1857 = *Werke*, pp. 88–144), were a more systematic way of handling any algebraic irrationality, not just the square root of a quotient. Riemann introduced the generalized theta function, which subsequently became the basic tool for solving the Jacobi inversion problem, namely,

$$\Theta(v_1, \ldots, v_p) = \left(\sum_{-\infty}^{\infty}\right)^p e^{(\Sigma_1^p)^2 a_{\mu\mu'} m_\mu m_{\mu'} + 2\Sigma_1^p v_\mu m_\mu},$$

where $(\Sigma_1^p)^2$ means $\Sigma_{\mu=1}^p \Sigma_{\mu'=1}^p$ and $(\Sigma_{-\infty}^{\infty})^p$ means $\Sigma_{m_p=-\infty}^{\infty} \cdots \Sigma_{m_1=-\infty}^{\infty}$.

It was probably clear that the Jacobi inversion problem could be solved using the theta function. The problem was how to carry out the details, i.e., how to adjust the coefficients $a_{\mu\mu'}$ so that the desired functions could be simply expressed in terms of the theta function. Here the Riemann surface enters the picture. Suppose an irreducible polynomial $f(x, y)$ gives rise to a system of Abelian integrals in which there are p independent integrals of first kind (bounded for all paths of integration). Riemann showed that the Riemann surface of the function defined by $f(x, y) = 0$ can be cut by $3p-1$ curves a_1, . . . , a_p, b_1, . . . , b_p, c_1, . . . , c_{p-1} in such a way that the period modulus (Cauchy's indice de périodicité) of the kth integral along c_j is 0 for all j and k. Along a_j it is πi if $j = k$, otherwise 0, while the modulus of the kth integral along b_j equals the modulus of the jth integral along b_k. The $2p$ periods of the inverse functions with this choice of period moduli are thus easily arranged in a p-by-$2p$ array (which Riemann did not write down)

$$
\begin{matrix}
\pi i & 0 & 0 & \cdots & 0 & \tau_{11} & \cdots & \tau_{1p} \\
0 & \pi i & 0 & \cdots & 0 & \tau_{21} & \cdots & \tau_{2p} \\
\cdot & & & & & & & \\
\cdot & & & & & & & \\
\cdot & & & & & & & \\
0 & 0 & 0 & \cdots & \pi i & \tau_{p1} & \cdots & \tau_{pp} \; .
\end{matrix}
$$

(For details of Riemann's discussion see his 1857, second part, §18–19 or *Werke*, pp. 129–130. A full explanation, with figures is given in Forsyth 1918, p. 402.)

Riemann said that the connection between his approach and that of Weierstrass was not clear, since Weierstrass had not yet given the details of the results he had announced. The connection was also not clear to Weierstrass, even after the appearance of Riemann's paper. Weierstrass was led by Riemann's work to start afresh in his efforts to clarify this complicated subject. For an account of the friendly competition between Riemann and Weierstrass on this subject, see Neuenschwander (1981, pp. 93–97).

3.11. Weierstrass' Later Work

Weierstrass neatly summarized the situation with regard to the Jacobi inversion problem and his own research in the introduction which he wrote for his lectures on Abelian functions when they were printed in his collected works. (Weierstrass died during the printing of this work and saw only the first 18 pages of it.):

. . . I now succeeded in representing the Abelian functions as quotients of two everywhere-converging power series. The numerators and denominators are polynomials in theta functions of p variables, and so I was led to the theta functions of arbitrarily many variables, whose form had previously been unknown to me.

However, Abel has extended the theorem which we have called Abel's theorem above to the integrals of algebraic functions arising from an arbitrary algebraic irrationality, not merely the hyperelliptic case. An inversion problem attaches to this extension of Abel's theorem also, and again the problem arises of representing the symmetric functions of the p pairs (x_ν, y_ν) as single-valued functions of p variables u_1, . . . , u_p in this more general case.

I presented a direct solution of this problem in a detailed report to the Berlin Academy in the summer of 1857. The manuscript, which had already gone to the printer, was withdrawn by me, however, when several weeks later Riemann published a work on the same problem which rested on entirely different foundations from mine and whose results could be shown to agree completely with mine only through further investigation. The proof of this result demanded certain investigations of a fundamentally algebraic nature whose execution turned out not entirely easy for me and claimed a great deal of time. But even after these difficulties were overcome, a thorough reworking of my report seemed necessary. Other tasks, equally important, which it is no longer of any interest to discuss, prevented me until the end of 1869 from giving that form to the solution of the general inversion problem which from then on was used in my lectures . . .

Weierstrass' collected works contain articles devoted to Abelian functions which are clearly later than the ones discussed thus far, but unfortunately some of them bear no indication as to the date of composition. One which does bear a date was published in the *Monatsberichte* of the Berlin Academy of Sciences at the time when Weierstrass says he had finally gotten the material organized to his satisfaction. In this article (1869) he investigated a general function of n complex variables having $2n$ periods and showed that many such functions even more general than those arising from Abelian integrals can yet be expressed as quotients of theta functions. He ended by asking whether any $2n$-periodic function of n complex variables could be so expressed, a question he finally answered affirmatively in a letter to Borchardt (1880 or *Werke* II, p. 133). A full exposition of Weierstrass' work on Abelian integrals can be found in a paper of W. Thimm (Behnke and Kopfermann 1966, pp. 123–154).

The "thorough reworking" Weierstrass spoke of resulted in a comprehensive and systematic treatment of the whole subject of Abelian integrals, now published as Volume IV of Weierstrass' *Werke*. These are the lectures which Kovalevskaya must have heard in 1872 and 1874. Before taking up the problem which Kovalevskaya herself worked on, it may be of interest to glance at the earlier parts of this course to give some idea of the final organization which Weierstrass gave to the material.

Where Riemann had used the Riemann surface as a primary tool, Weierstrass took the concept which he called a "structure" (Gebilde), defined as the set of all pairs of complex numbers satisfying $f(x, y) = 0$, together with certain ideal points at infinity needed to close up the structure. (As usual, f is an irreducible polynomial.) At a point (x_0, y_0) where the partial derivatives of f are not both zero, one can find two analytic functions $\phi(t)$ and $\psi(t)$ such that $\phi(0) = x_0$, $\psi(0) = y_0$, and every pair (x, y) satisfying the equation and sufficiently close to (x_0, y_0) equals $(\phi(t), \psi(t))$ for exactly one value of t close to 0. The local properties of the structure can then be studied using this pair of analytic functions. Weierstrass called such a pair an *element* of the structure. (Weierstrass' use of elements differed somewhat from the modern use of charts and manifolds.) He called two elements $(\phi(t), \psi(t))$ and $(\phi_1(t),$

$\psi_1(t)$) equivalent if they have the same midpoint (i.e., the same value for $t = 0$) and, in his words, "in a certain neighborhood of the midpoint each point of one is also a point of the other." [I think this means there are *arbitrarily small* neighborhoods U and U_1 of 0 such that the image of U under the mapping $t \rightarrow (\phi(t), \psi(t))$ is the same as that of U_1 under $t \rightarrow (\phi(t), \psi(t))$.] Weierstrass showed that two elements are equivalent if and only if there is an analytic function $g(t)$ with $g(0) = 0$ such that $(\phi(t), \psi(t)) = [\phi_1(g(t)), \psi_1(g(t))]$.

Perhaps an example may make this clearer (not one given by Weierstrass, however). If $f(x, y) = y^2 - x^3$, a neighborhood of $(0,0)$ can be covered by the image of a small disk about 0 under the mapping $t \rightarrow (t^2, t^3)$. In this case only one element is needed to describe the neighborhood of $(0,0)$. On the other hand, if $f(x, y) = y^2 - x^2 - x^3$, one can define two essentially different elements around $(0,0)$ as follows:

$$\phi(t) = t(t+2),$$
$$\psi(t) = t(t+2)(t+1)$$

and

$$\phi_1(t) = t(t+2),$$
$$\psi_1(t) = -t(t+2)(t+1).$$

For small values of t these two elements have only the point $(0,0)$ in common. They are therefore not equivalent. Weierstrass showed how to find a finite set of inequivalent elements to cover a neighborhood of any point. If such a set consists of λ elements, the point is called a λ-fold point.

Weierstrass next set out to find the simplest function on a structure having prescribed poles. The problem would be simple if there existed a function $F(x, y; x_0, y_0)$ having a simple pole at (x_0, y_0) and no other singularities. For example for the structure given by the polynomial $f(x, y) = y - x$, one can take $f(x, y; x_0, y_0) = (x - x_0)^{-1}$. Then any rational function can be represented using a finite combination of such functions; the technique for doing so is simply partial fractions. In general if such a function exists, the constants $C_0, \ldots, C_\nu$ can be chosen so that the function

$$C_0 + C_1 F(x, y; x_0, y_0) + \cdots + C_\nu F(x, y; x_\nu, y_\nu)$$

has any prescribed zeros $(x^1, y^1), \ldots, (x^\nu, y^\nu)$ and assumes a prescribed value at any desired point. For one need only solve the system

$$1 + u_1 F(x^j, y^j; x_1, y_1) + \cdots + u_\nu F(x^j, y^j; x_\nu, y_\nu) = 0,$$

$j = 1, 2, \ldots, \nu$, for $u_1, u_2, \ldots, u_\nu$. Then the function

$$G(x, y) = 1 + u_1 F(x, y; x_1, y_1) + \cdots + u_\nu F(x, y; x_\nu, y_\nu)$$

has the prescribed zeros and can be multiplied by a constant to give the desired value at any other point.

As an example of this technique, Weierstrass gave the polynomial $f(x, y) = Ax^2 + Bxy + Cy^2 + Dx + Ey + F$. If (x_0, y_0) is any point for which $f(x_0, y_0) = 0$, simply set

$$F(x, y; x_0, y_0) = \frac{Bx_0 + E + C(y + y_0)}{x - x_0}.$$

There are only two values of y corresponding to each value of x. For $x = x_0$, F becomes infinite for one of the two y values, but not for the other. Hence F has only the simple pole at one point of the structure defined by $f(x, y) = 0$. It follows that on any structure defined by an equation of second degree we can easily construct a rational function having a prescribed set of ν poles, ν zeros, and a prescribed value at any additional point.

In general, however, a function as nice as the one just constructed does not exist. The best that can be done is to minimize the number of "excess" poles. Weierstrass defined the rank of a structure to be p provided $p + 1$ is the smallest number such that if any $p+1$ points of the structure are chosen, there exists a rational function of x and y having simple poles at precisely these points. It turns out that p is equal to the number of independent Abelian integrals of first kind, hence equal to the number of variables in the Jacobi inversion problem, as well as the number of integrals to which any sum of integrals can be reduced by Abel's theorem. [Also, though there is no need for this fact here, the rank of the structure is equal to the genus of the Riemann surface defined by $f(x, y) = 0$.]

For a structure of rank p Weierstrass showed how to construct a rational function $F(x, y; x', y')$ with a simple pole at (x', y') and p other poles at specified points (a_j, b_j). After the function F is constructed, the function

$$H(x, y; x'y') = \frac{F(x, y; x', y') - F(a_0, b_0; x', y')}{f(x', y')_2}$$

becomes a fundamental tool (the subscript 2 denotes partial differentiation). It has the same poles as F and vanishes at (a_0, b_0). If an element $x = \phi(t)$, $y = \psi(t)$ is chosen at (a_j, b_j), $j = 1, 2, \ldots, p$, the function $H(x, y; x', y')$ can be expanded in a series of powers of t, starting with t^{-1}:

$$H = \sum_{k=-1}^{\infty} c_k t^k.$$

Then the residue c_{-1} depends on (x', y') and j and can be considered a function $H(x', y')_j$. It turns out that for any element $(\phi(t), \psi(t))$ the function $H(\phi(t), \psi(t))\phi'(t)$ is analytic at $t = 0$. Consequently the indefinite integrals $\int H(x, y)_j \, dx$ are everywhere finite, i.e., they are integrals of first kind. Since these integrals are independent, every integral of first kind is a linear combination of them. They can therefore be used to formulate the Jacobi inversion problem. With this formulation the Jacobi inversion problem becomes tractable. For details, however, the reader is referred to Weierstrass' *Werke*. It is at long last time to examine Kovalevskaya's contribution to Abelian integrals.

3.12. Kovalevskaya's Paper

The problem Weierstrass assigned to Kovalevskaya involves Abelian integrals, though it is not one of the problems of central importance to the theory. He asked her to investigate cases of degeneracy, in which Abelian integrals reduce to elementary or elliptic integrals. Occasional papers devoted entirely to this subject were seen throughout the nineteenth century, for instance, Richelot (1846). The subject also came up incidentally in papers primarily devoted to hyperelliptic integrals; for instance, Jacobi (1832c) showed that certain hyperelliptic integrals of first kind reduce to elliptic integrals when the basic equation defining the integrals is $y^2 = x(1 - x)(1 + \kappa x)(1 + \lambda x) \cdot (1 - \kappa\lambda x)$. [Integrals based on this equation arise in practice in determining the capacitance of a condenser consisting of one square prism inside another. Even after the reduction, however, the elliptic integrals which result are too cumbersome to be of practical use (cf. Bowman 1953, pp. 99–108).]

Kovalevskaya's old teacher Leo Koenigsberger wrote several papers on Abelian integrals and theta functions during the late 1860s. In one of these (1865) he communicated Weierstrass' notation for some theta functions, a notation which Kovalevskaya used in her paper. This notation is as follows. Starting with Riemann's theta function $\Theta(v_1, \ldots, v_p)$ defined above, let

$$\tau_a = n_1 \tau_{a1} + n_2 \tau_{a2} + \cdots + n_p \tau_{ap},$$

and set

$$\Theta(v_1, \ldots, v_p; n_1, \ldots, n_p) = \Theta(v_1 + \tau_1, \ldots, v_p + \tau_p)e^{\Sigma_a n_a(2v_a + \tau_a)\pi i}.$$

Koenigsberger then introduced a subscript notation for translation of this theta function by half-periods, namely,

$$\Theta(v_1, \ldots, v_p)_{\lambda\mu} = \Theta(v_1 + \tfrac{1}{2}m_1^\nu, \ldots, v_p + \tfrac{1}{2}m_p^\nu; \tfrac{1}{2}n_1^\nu, \ldots, \tfrac{1}{2}n_p^\nu)$$

when $m_a^\nu \equiv m_a^\lambda + m_a^\mu$ mod 2 and $n_a^\nu \equiv n_a^\lambda + n_a^\mu$ mod 2 for all a. The meaning of $\Theta(v_1, \ldots, v_p)_{\lambda\mu\nu}$, etc., can be inferred easily.

One of Koenigsberger's papers (1867) investigated the question of which Abelian integrals of rank 2 reduce to elliptic integrals via a transformation of periods of degree 2, i.e., $k = 2$ in the transformation given by Hermite. Kovalevskaya's assignment was simple: Do the same for rank 3, keeping $k = 2$. As in her paper on partial differential equations, she began by explaining what she had learned from Weierstrass that a reader not in contact with him would not be likely to know, then set out her own ideas.

The starting point for her paper (1884) is a corollary of a theorem of

Abel (*Oeuvres* I, p. 546). The corollary says that if an Abelian integral
reduces to elliptic integrals, then there exists an integral of the form

$$\int \frac{R(s)\,ds}{\sqrt{\phi(s)}},$$

to which the original integral can be reduced and such that s, $R(s)$, $\phi(s)$, and
$\sqrt{\phi(s)}$ are all rational functions of the original variables x and y, ϕ being a
polynomial of degree three or four in s. Weierstrass, said Kovalevskaya, had
deduced as a corollary that if there is some Abelian integral $\int R(x,y)\,dx$
which reduces to an elliptic integral, then there is an Abelian integral of first
kind depending on the same equation between x and y which also reduces to
an elliptic integral. For, by the theorem just quoted, it is possible to find s and
$\phi(s)$ such that $\sqrt{\phi(s)}$ is a rational function of x and y. Then the elliptic
integral $\int (1/\sqrt{\phi(s)})\,ds$ is an Abelian integral in terms of x and y and at the
same time it is bounded, hence of first kind. (Evidently this is the argument;
Kovalevskaya's comments seem rather cryptic.) The question of degeneracy
thus becomes: Are there Abelian integrals of first kind associated with
$f(x,y) = 0$ which reduce to elliptic integrals?

Weierstrass changed the question around by taking p independent integrals of first kind, say $u_1, \ldots, u_p$, and asking what conditions on the
coefficients $c_1, \ldots, c_p$ cause $u = c_1 u_1 + \cdots + c_p u_p$ to reduce to an elliptic integral. His answer to this question was essentially that if such an integral
exists, it must lead to a transformation of periods of a very special type. His
argument, as given by Kovalevskaya, is as follows:

Suppose the integrals $u_1, \ldots, u_p$ have periods $(1,0,0,\ldots,0)$,
$(0,1,0,\ldots,0), \ldots, (0,0,0,\ldots,1), (\tau_{11},\ldots,\tau_{p1}), \ldots, (\tau_{1p},\ldots,$
$\tau_{pp})$, that is, the integrals taken around $2p$ suitably chosen curves give these
$2p$ sets of values. Then u must have the $2p$ periods $c_1, \ldots, c_p$,
$c_1\tau_{11} + \cdots + c_p\tau_{p1}, \ldots, c_1\tau_{1p} + \cdots + c_p\tau_{pp}$. But u, being an elliptic
integral, has only two independent periods, say 2ω and $2\omega'$. Hence there must
be integers m_j, n_j, m_j', n_j' such that

$$c_1 = 2n_1\omega + 2n_1'\omega',$$
$$\vdots$$
$$c_p = 2n_p\omega + 2n_p'\omega',$$
$$c_1\tau_{11} + \cdots + c_p\tau_{p1} = -2(m_1\omega + m_1'\omega')$$
$$\vdots$$
$$c_1\tau_{1p} + \cdots + c_p\tau_{pp} = -2(m_p\omega + m_p'\omega').$$

Using the integers m_j, n_j, m_j', n_j' to start, there are in general many different
ways to construct a transformation of periods so as to obtain 2ω and $2\omega'$
among the transformed periods. [When $p = 3$, however, there is only one

way, (cf. Baker 1897, p. 676).] The degree of the transformation is the integer $k = m_1 n_1' + \cdots + m_p n_p' - (m_1' n_1 + \cdots + m_p' n_p)$, which must be positive. When such a transformation is carried out, the new periods which result are $(1, 0, 0, \ldots, 0), \ldots, (0, 0, 0, \ldots, 1) (\tau_{11}', \ldots, \tau_{p1}'), \ldots, (\tau_{1p}', \ldots, \tau_{pp}')$ and the equations just written for the coefficients imply that $\tau_{12}' = \tau_{13}' = \cdots = \tau_{1p}' = 0$. That is to say, the theta function constructed taking $a_{\mu\mu'} = \tau_{\mu\mu'}'$ (cf., Section 3.10) breaks up into the product of a theta function of v_1 and a theta function of $v_2, \ldots, v_p$. This result is Weierstrass' transcendental condition for degeneracy: Among the theta functions associated with the Abelian integrals in question, there must be one which has this simple form.

Kovalevskaya next presented a theorem of Weierstrass which showed that if a transformation of degree k, such as the one above, leads to periods of the given form, then there is a transformation of degree one which leads to periods with $\tau_{13}' = \cdots = \tau_{1p}' = 0$ while $\tau_{12}' = r/k$, where r is a non-negative integer less than k. For the case $k = 2$, $p = 3$, it should then be possible to work Weierstrass' transcendental criterion into something more usable. That is what Kovalevskaya did. Describing her own approach to the problem she said

> I then noticed that on the basis of this proposition the relations being sought could also be obtained without the theory of transformations by taking as a basis the following theorem deduced from the lectures of Herr Weierstrass on Abelian functions.
>
> "Let $x, y, u_1, \ldots, u_p, \tau_{11}, \ldots, \tau_{pp}$ have the same meanings as before, and let $u_1', \ldots, u_p'$ be the values of $u_1, \ldots, u_p$ at a different point. Among the theta functions with moduli $\tau_{\alpha\beta}$ choose any two odd ones $\Theta(v_1, \ldots, v_p)_\lambda$ and $\Theta(v_1, \ldots, v_p)_\mu$. Then if we set
>
> $$\frac{du_\alpha}{dx} = \overline{H}(x, y)_\alpha, \quad \frac{du_\alpha'}{dx} = \overline{H}(x', y')_\alpha,$$
>
> $$\left\{ \frac{\Theta(u_1 - u_1', \ldots, u_p - u_p')_\lambda}{\Theta(u_1 - u_1', \ldots, u_p - u_p')_\mu} \right\}^2 = \sum_\alpha \frac{\Theta_\lambda^{(\alpha)} \overline{H}(x', y')_\alpha}{\Theta_\mu^{(\alpha)} \overline{H}(x', y')_\alpha} \sum_\alpha \frac{\Theta_\lambda^{(\alpha)} \overline{H}(x, y)_\alpha}{\Theta_\mu^{(\alpha)} \overline{H}(x, y)_\alpha}$$
>
> where $\Theta_\gamma^{(\alpha)}$ and $\Theta_\mu^{(\alpha)}$ denote the values of $\partial \Theta(v_1, \ldots, v_p)_\gamma / \partial v_\alpha$ and $\partial \Theta(v_1, \ldots, v_p)_\mu / \partial v_\alpha$ when $v_1 = 0, \ldots, v_p = 0$."

The subscript notation used here is the one introduced by Koenigsberger. Since Kovalevskaya's formula connects the theta functions with rational functions of x and y, any equation satisfied by the theta functions will imply an equation which must be satisfied by x and y. Hence the transcendental equations will become algebraic. The progression of the argument seems to be: degeneracy of an integral implies relations for the theta functions, which in turn imply restrictions on x and y. The rest of the paper is an execution of this program.

With the rank fixed at 3 and $k = 2$, Weierstrass' proposition implies that if a degenerate Abelian integral exists, then some theta function associated with the given polynomial equation must be constructed using as parameters the following array

$$
\begin{array}{ccc}
\tau_{11} & \tfrac{1}{2} & 0 \\[4pt]
\tfrac{1}{2} & \tau_{22} & \tau_{23} \\[4pt]
0 & \tau_{32} & \tau_{33}
\end{array}
$$

with $\tau_{23} = \tau_{32}$. Of the many theta functions derived from this one by doing the half-period translations which Koenigsberger denoted with subscripts, Kovalevskaya selected several for which the following convenient relation held:

$$
g_1 \Theta(v_1, v_2, v_3)_5 \Theta(v_1, v_2, v_3)_{125} + g_2 \Theta(v_1, v_2, v_3)_{345} \Theta(v_1, v_2, v_3)_{06}
$$
$$
+ g_3 \Theta(v_1, v_2, v_3)_{46} \Theta(v_1, v_2, v_3)_{035} + g_4 \Theta(v_1, v_2, v_3)_3 \Theta(v_1, v_2, v_3)_{123} = 0.
$$

By differentiating this equation and applying other known relations, Kovalevskaya obtained an elegantly symmetrical equation in x and y:

$$
h_1 \sqrt{\xi_1 \xi_1'} + h_2 \sqrt{\xi_2 \xi_2'} + h_3 \sqrt{\xi_3 \xi_3'} = 0. \tag{4}
$$

Here h_1, h_2, and h_3 are constants, and

$$
\xi_1 = \Theta_5^{(1)} \overline{H}(x, y)_1 + \Theta_5^{(2)} \overline{H}(x, y)_2 + \Theta_5^{(3)} \overline{H}(x, y)_3.
$$

Similar expressions hold for ξ_2, ξ_3, ξ_1', ξ_2', ξ_3', with $\Theta_5^{(j)}$ replaced by $\Theta_{345}^{(j)}$, $\Theta_{46}^{(j)}$, $\Theta_{125}^{(j)}$, $\Theta_{06}^{(j)}$, and $\Theta_{035}^{(j)}$, respectively. Thus the ξ_j' are homogeneous linear functions of ξ_1, ξ_2, and ξ_3. Now ξ_1, ξ_2, and ξ_3 can be regarded as homogeneous coordinates of a point. [If the point (u, v, w) is identified with $(u/w, v/w)$ then u, v, and w are homogeneous coordinates of the point.] Thus equation (4) represents a curve when ξ_j' are expressed in terms of ξ_1, ξ_2, and ξ_3 and the latter are regarded as homogeneous coordinates. The six "lines" $\xi_j = 0$, $\xi_j' = 0, j = 1, 2, 3$, are double tangents of this curve, and the assumptions about the theta function imply that the first four of these intersect in a single point corresponding to $\overline{H}_2 = 0 = \overline{H}_3$.

Then, remarking that $H(x, y)_j, j = 1, 2, 3$, could be chosen so as to give three independent integrals of first kind, and setting $X_j = H(x, y)_j$, she showed that the variables in the symmetric equation above are all homogeneous linear functions of X_1, X_2, X_3. Thus, when these variables are replaced by their expressions in terms of the X_j's and the radicals are cleared, the result is a homogeneous equation of degree 4:

$$
F(X_1, X_2, X_3) = 0.
$$

(To carry out this process, one need only eliminate x and y from the three equations

$$
f(x, y) = 0, \qquad X_1 H(x, y)_2 = X_2 H(x, y)_1, \qquad X_2 H(x, y)_3 = X_3 H(x, y)_2.
$$

Then, except for one special situation which will be discussed below, Kovalevskaya had the following theorem:

If y is an algebraic function of x of rank 3, then the equation which holds between x and y can be transformed (in infinitely many ways) into a homogeneous equation of degree four, $F = 0$, among three quantities X_1, X_2, X_3 which are rational functions of x and y. Moreover this can always be done so that the coefficients of this equation and of the expressions X are formed rationally from those of the given equation. The equation $F = 0$ is irreducible and represents a curve of order four without double points.

The exceptional case occurs when x and y are not rational functions of

$$\xi = \frac{H(x,y)_1}{H(x,y)_3}$$

and

$$\eta = \frac{H(x,y)_2}{H(x,y)_3}.$$

In that case the equation $F(\xi, \eta, 1) = 0$ is not equivalent to the given equation. Kovalevskaya claimed that this case occurs only when there are at most two values of η satisfying $F(\xi, \eta, 1) = 0$ for each value of ξ. (This claim may be justified by referring to the process by which the new equation is derived from the old.)

The second and more important part of Kovalevskaya's theorem is the following:

Among the double tangents of this equation four must coincide, i.e., their eight points of tangency must lie on a single conic section; they intersect in *one* point if among the Abelian integrals depending on y there is at least one which can be reduced to an elliptic integral by a transformation of the type under consideration ($k = 2$).

It can be seen that Kovalevskaya's condition is a purely algebraic one, phrased in terms of the tangents to the curve. For this reason it is preferable to the transcendental conditions given by Weierstrass.

For the converse of this result Kovalevskaya produced an elegant series of changes of variable, converting the equation $F(X_1, X_2, X_3) = 0$ into $G(p,q,r) = 0$, where p, q, and r are homogeneous linear functions of X_1, X_2, X_3 and $G(p,q,r) = R_0 r^4 + R_2 r^2 + R_4$, each R_j being a homogeneous function of p and q of degree j. Then if $r = c_1 X_1 + c_2 X_2 + c_3 X_3$ and F satisfies the condition on its tangents enunciated above, it follows that

$$\int \frac{c_1 X_1 + c_2 X_2 + c_3 X_3}{\partial F / \partial X_3} (X_1 dX_2 - X_2 dX_1),$$

which is an Abelian integral when written in terms of x and y, is equal to $\int d\xi / \sqrt{R(\xi)}$, where $\xi = q/r$, $R(\xi) = R_2^2(1, \xi) - R_4(1, \xi)$. Since

$$\sqrt{R(\xi)} = R_0 \frac{r^2}{p^2} + \frac{R_2(p,q)}{p^2},$$

both ξ and $\sqrt{R(\xi)}$ are rational functions of x and y, in accordance with the theorem of Abel quoted at the beginning of the paper.

By analyzing the construction in more detail, she showed that the coefficients of G determine how many such reducible integrals there are. In fact, if $G(p, q, r) = r^4 + (L_1 p^2 + L_2 pq + L_3 q^2)r^2 + M_1 p^4 + M_2 p^3 q + M_3 p^2 q^2 + M_4 pq^3 + M_5 q^4$, then normally there are only two independent integrals of first kind which reduce to elliptic. The exception occurs when both L_2 and M_2 vanish; in that case there are three independent integrals of first kind which reduce.

Kovalevskaya concluded her paper with an investigation of the exceptional case where for each value of ξ there are at most two values of η satisfying $F(\xi, \eta, 1) = 0$. Since this assumption means F is essentially of degree 2 in η, the integrals are hyperelliptic, i.e., they come from the equation $\eta^2 = g(\xi)$, where the degree of g must be 7 or 8, since the rank is 3. An integral of the first kind is then of the form

$$\int \frac{c_0 + c_1 \xi + c_2 \xi^2}{\sqrt{g(\xi)}} \, d\xi.$$

Kovalevskaya's result, stated without proof, is as follows: If one of these integrals reduces to an elliptic integral by a transformation of degree 2, then the roots of g (say $a_0, \ldots, a_7$) can be so labeled that

$$1 = \frac{(a_0 - a_1)(a_2 - a_7)(a_3 - a_4)(a_5 - a_6)}{(a_0 - a_7)(a_1 - a_2)(a_3 - a_6)(a_4 - a_5)}$$

and

$$1 = \frac{(a_0 - a_5)(a_1 - a_6)(a_2 - a_3)(a_4 - a_7)}{(a_0 - a_3)(a_1 - a_4)(a_2 - a_5)(a_6 - a_7)}.$$

Conversely, if these two conditions hold and in addition

$$1 = \frac{(a_0 - a_4)(a_1 - a_3)(a_2 - a_6)(a_5 - a_7)}{(a_0 - a_6)(a_1 - a_5)(a_2 - a_4)(a_3 - a_7)},$$

then there exist three linearly independent reducible integrals of first kind.

3.13. Evaluation

This paper was considered a minor result at the time. Weierstrass gave an evaluation of it in the letter he wrote to Fuchs requesting a degree for Kovalevskaya (Mittag-Leffler 1923f, p. 249). He pointed out that obtaining the solution of this problem required less inventiveness than the paper on partial differential equations, but nevertheless did demonstrate a solid understanding of Abelian functions. Such a demonstration was proof of a high level of mathematical competence, just the thing most needed by the first woman Ph.D. in mathematics, who was unfortunately fated to be proving herself more than any male Ph.D. would have to do.

The principal achievement in this paper, as already pointed out, was the replacing of a transcendental criterion for degeneracy with an algebraic one. To Weierstrass this result was very important. As he wrote to H. A. Schwarz on 3 October 1875 (*Werke* II, p. 235):

> . . .The more I ponder the principles of function theory – and I do this constantly – the firmer becomes my conviction that the latter must be built on the basis of algebraic facts, and that it is therefore not the right way to proceed when contrariwise the "transcendental" – to use a short expression – is used to establish simple and basic algebraic theorems . . . .

Although the topic of degeneracy of Abelian integrals was not one of the central parts of the theory, it nevertheless received attention from many mathematicians toward the end of the nineteenth century. For example, it occupies the last chapter of Krazer's book on theta functions (1903) and the last chapter of Baker's treatise on Abelian functions (1897). According to Baker (1897, p. 663) the usual procedure was to start with an elliptic integral and transform it into an Abelian integral.

Although several papers on this topic appeared during the 1870s, none of them duplicated Kovalevskaya's work, despite the fact that she had not published it. The closest approach to her work was an 1884 paper by Poincaré in which it was proved that if a system of Abelian integrals of rank p contains more than p integrals of first kind which reduce to elliptic integrals, then it contains infinitely many.

The fact that no one duplicated this work during the next 6 years made it possible for Kovalevskaya to use this paper to get reintroduced to the world of mathematics in 1880 and to publish it in the *Acta Mathematica* in 1884 as she began her career as dozent in Stockholm.

CHAPTER 4

The Shape of Saturn's Rings

4.1. Introduction

Kovalevskaya's paper (1885b) on the shape of Saturn's rings was written while she was working with Weierstrass, but quite likely on her own initiative, since Weierstrass does not seem to have touched on the topic in either his published work or in his lectures. The title, "Zusätze und Bemerkungen zu Laplace's Untersuchung über die Gestalt der Saturnringe," shows the context of the work. Laplace (1799) had studied the shape and stability of what was then believed to be Saturn's one ring, though he did mention (1799, II, p. 163) some evidence that there was more than one ring. Laplace assumed that the ring had some kind of continuous structure. However Maxwell (1859) had shown that it was very unlikely that the rings could have any continuous structure such as Laplace's work evidently postulated. It would seem to follow that further work on the physical model used by Laplace would not be worthwhile. Kovalevskaya therefore phrased her results as a commentary on Laplace.

The plan of this chapter is as follows. First Laplace's work on the shape of the ring(s) will be discussed, with special attention to the mathematically most difficult part of the problem, computing the potential of a ring. This discussion occupies Sections 4.2–4.5. The remainder of the chapter is devoted to the works of other mathematicians on the subject, which means essentially Kovalevskaya. The shape of the rings, as opposed to their stability, does not seem to have interested many people. Consequently Kovalevskaya had the field pretty much to herself.

4.2. Laplace's Work: Physical Assumptions

While remaining noncommittal about the physical composition of Saturn's ring, Laplace (1799, p. 155) expressed the opinion that the continued existence of the ring could not depend on intermolecular forces. He therefore thought it reasonable that the external forces acting on the ring must somehow not put any strain on the ring. He concluded that a thin layer of liquid on the surface of the ring under such conditions would be in equilibrium. (It seems clear from his words that the thin layer of liquid was a fiction to be used to determine the shape of the ring, rather than a physical reality for Laplace.) The mathematical formulation of the conditions of equilibrium for such a layer of liquid would then determine the shape of a cross section of the ring.

The problem, then, is to determine a shape such that the forces acting at each point of the surface of the ring, namely the gravitational attraction of

Saturn and of the ring itself, produce at that point a resultant force consisting of two components: (1) a centripetal force to hold the point in circular orbit as the ring rotates; (2) a force normal to the surface of the ring, which is therefore nullified by the ring itself.

4.3. Laplace's Work: Mathematical Assumptions

In his investigation Laplace assumed that the figure of the ring is generated by revolving a plane curve about a line in the plane of the curve but exterior to the curve. As the work progressed, he further assumed that the curve is symmetric about a line perpendicular to the axis of rotation. In the end, before proceeding to computations, he assumed that the curve is in fact an ellipse. Thus by the end of his work, determining the shape of the curve meant merely determining a single number, namely the eccentricity of the ellipse. Since it is a trivial matter to calculate both the centripetal force needed to hold a particle in orbit and the gravitational attraction of Saturn, the principal difficulty of the problem arises in calculating the gravitational attraction of the ring itself. This part of the problem will occupy most of the exposition which follows. Consequently, it seems desirable to first give an outline of the procedure by which the computation is carried out.

Laplace first derived the general analytical form which the potential of the ring must have. From this form it appeared that the potential at every point could be found if the potential were known at the points of one axis. For points along this axis the function whose integral gives the potential can be reasonably approximated by functions whose integrals are directly computable. In both stages of this work, deriving the general form of the potential and computing its value along one axis, Laplace made use of an approximation whose validity and improvement provided the impetus for Kovalevskaya's work.

4.4. The Potential of the Ring

In an earlier part of the *Mécanique Céleste* (Book II, no. 11) Laplace had shown that if a spheroid attracts a point mass, then "V being the sum of the molecules of the spheroid divided by their distances from the point," we have the now-famous Laplacian equation

$$\frac{\partial^2 V}{\partial x^2} + \frac{\partial^2 V}{\partial y^2} + \frac{\partial^2 V}{\partial z^2} = 0.$$

The function V is what is now called the gravitational potential at the point (x,y,z), i.e.,

$$V(x, y, z) = \int\int\int ((x - u)^2 + (y - v)^2 + (z - w)^2)^{-1/2}\, \rho(u, v, w)\, du\, dv\, dw,$$

where ρ is the density of the spheroid and the integral extends over the region occupied by the spheroid.

Since in the present case the body whose potential is being calculated is a ring, V is best expressed in cylindrical coordinates. It is then independent of the angular coordinate, and Laplace's equation becomes

$$\frac{1}{r}\frac{\partial V}{\partial r} + \frac{\partial^2 V}{\partial r^2} + \frac{\partial^2 V}{\partial z^2} = 0.$$

(In this coordinate system, the origin is the center of Saturn and the z axis is the axis of the rotation which produces the ring.) To simplify this equation Laplace assumed the ring to be rather far from the planet in comparison with its dimensions. More precisely, he took a to be the distance from the center of Saturn to the center of the horizontal bisector of the generating curve, wrote $r = u + a$, and assumed u and z very small compared with a for points on the surface of the ring (see Figure 4-1).

The first term in the cylindrical version of Laplace's equation can then be neglected and the equation becomes

$$\frac{\partial^2 V}{\partial u^2} + \frac{\partial^2 V}{\partial z^2} = 0.$$

For convenience the approximation just made will be referred to as "approximation A." It leads to an important geometrical interpretation. Whereas the original equation reflected the independence of V from the angular coordinate, the new equation is precisely the two-variable Laplace equation in rectangular coordinates. Consequently it can be interpreted as representing the potential of a body whose symmetry is linear rather than rotational, i.e., V is independent of a third rectangular coordinate. Thus, in effect, instead of rotating the

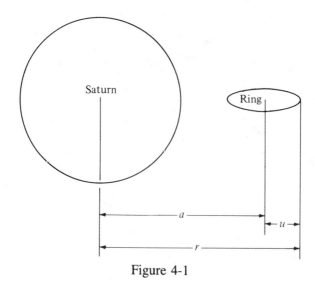

Figure 4-1

curve to produce a ring, we have translated it perpendicular to its plane to produce an infinitely long cylinder.

Laplace was familiar with d'Alembert's solution of the equation of the vibrating string (cf. Kline 1972, p. 503)

$$c^2 \frac{\partial^2 s}{\partial u^2} = \frac{\partial^2 s}{\partial z^2}.$$

By letting $x = u + cz$, $y = u - cz$, d'Alembert converted this equation into the equation $\partial^2 s / \partial y \partial x = 0$, whose general solution is obviously $s = \phi(x) + \psi(y)$, where ϕ and ψ are arbitrary functions. Since Laplace's equation can be regarded as the special case $c^2 = -1$, Laplace wrote the general solution of his equation as

$$V(u, z) = \phi(u + z\sqrt{-1}) + \psi(u - z\sqrt{-1}).$$

(In this connection it should be noted that Laplace undoubtedly thought of the symbols ϕ and ψ as representing specific algebraic formulas, for which imaginary numbers have at least a formal sense.) By requiring V to be real valued and taking account of the geometrical symmetry of V, Laplace deduced that V must have the form

$$V(u, z) = f(u + z\sqrt{-1}) + f(u - z\sqrt{-1}).$$

This formula completes the first stage of the program outlined in the introduction to this chapter. Formally the function f is determined by the equation $f(u) = \frac{1}{2} V(u, 0)$, and so V is determined by its values for $z = 0$. (Laplace seems to be proceeding formally here. Modern complex analysis justifies him in doing so, since anything he would have regarded as a function is determined by its values on the real line.)

Before passing to the explicit computation of V, it is worthwhile to discuss approximation A in more detail. It seems reasonable on physical grounds that the gravitational attraction of a large ring at a point on the ring does not differ greatly from that of a large cylinder with the same cross section. For, in a neighborhood of the point the two figures are practically the same, and points farther away have little influence on the attraction. The approximation is therefore a reasonable one. On the other hand, since it was not accompanied by any estimate of the order of magnitude of the error thus introduced, one can see a need to investigate the matter further. Laplace contented himself with sketching a power series expansion whose partial sums would give better and better approximations to the true potential.

Having established that it is sufficient to calculate the potential at points $(u, 0)$, Laplace proceeded to perform these calculations. Assuming that the equation of the generating curve is $y^2 = \phi^2(x)$, $-k \le x \le k$, one finds the potential at the point $(u, 0)$ to be

$$\int_{-k}^{k} \int_{-\phi(x)}^{\phi(x)} \int_{-\pi}^{\pi} (a + x) \tag{1}$$

$$\times [(a + u)^2 - 2(a + u)(a + x)\cos \omega + (a + x)^2 + y^2]^{-1/2} \, d\omega \, dy \, dx.$$

Approximation A, which amounts to the assumption that a is very large, becomes embarrassing at this point, since this integral tends to infinity with a. [It contains the integral of $(2 - 2 \cos \omega)^{1/2}$ for a infinite.] We are really interested in the *force* rather than the potential, however; that is, we want to calculate $\partial V/\partial u$, and the integral which gives that quantity is not infinite. Laplace took approximation A to mean that the following approximate equation could be used:

$$\int_{-a\pi}^{a\pi} \left(1 + \frac{x}{a}\right)((u - x)^2 + t^2 + y^2 + a^{-1}(x - u)t^2 + a^{-2}uxt^2)^{-1/2} \, dt$$

(2)

$$= \int_{-\infty}^{\infty} ((u - x)^2 + y^2 + t^2)^{-1/2} \, dt.$$

The left-hand side of (2) follows from (1) when $\cos \omega$ is approximated by $1 - \frac{1}{2}\omega^2$ and the change of variable $\omega = t/a$ is made. [Laplace did not actually write down equation (2), but he did give its right-hand side as the limit of the integral (1) when a tends to infinity. Both these integrals for the potential are infinite, as mentioned above. However, the partial derivative on u is *not* infinite, and it is this partial derivative which we need for the force. When the exponent $-\frac{1}{2}$ is decreased to $-\frac{3}{2}$, which is the result of differentiating on u inside the integral sign, both sides of (2) become finite, and one can *nowadays* justify the equation using the monotonic convergence theorem when $0 > x > u$. Laplace, of course, operated purely formally.]

Starting from this replacement it is easy to compute the u-component of the gravitational attraction of the ring as

$$\frac{\partial V}{\partial u} = -4 \int_{-k}^{k} \arctan \frac{\phi(x)}{u - x} \, dx$$

when $z = 0$. To go beyond this last equation it is necessary to know the function $\phi(x)$. At this point Laplace introduced the assumption that the generating curve is an ellipse, i.e., that

$$\phi(x) = \frac{(k^2 - x^2)^{1/2}}{\lambda},$$

where λ is the ratio of the major and minor axes of the ellipse. He then laconically gives the attraction of the ring as

$$-\frac{\partial V}{\partial u} = \frac{4\pi\lambda}{\lambda^2 - 1} \left\{ u - \left(u^2 - k^2\frac{\lambda^2 - 1}{\lambda^2}\right)^{1/2} \right\}.$$

[The verification seems complicated. I had to make repeated use of fractional-linear changes of variable to carry it out. Bowditch (1829, pp. 503–504) indicates another possible justification.]

4.5. Final Computations

The remaining work is straightforward. As already noted, $f'(u) = \frac{1}{2}(\partial V/\partial u)(u, 0)$. Then

$$\frac{\partial V}{\partial u}(u, z) = f'(u + z\sqrt{-1}) + f'(u - z\sqrt{-1}),$$

so that for points on the surface of the ring, where k^2 can be replaced by $u^2 + \lambda^2 z^2$, a suitable choice of square roots gives

$$\frac{\partial V}{\partial u} = -\frac{4\pi u}{\lambda + 1}.$$

A similar argument gives the z-component of the force as $\partial V/\partial z = -4\pi\lambda z/(\lambda + 1)$. Not having vector notation, Laplace of course phrased these assertions differently. He said that the attraction of the ring was

$$\frac{-4\pi u}{\lambda + 1} \, du - \frac{4\pi\lambda z}{\lambda + 1} \, dz.$$

Similarly, if S is the mass of Saturn and units of measure are suitably chosen, the gravitational attraction of Saturn at a point (u, z) is, in Laplace's terms $S((a + u)^2 + z^2)^{-1}$ in the direction of $-d((a + u)^2 + z^2)^{1/2}$, which works out to an attraction equal to $-S((a + u)^2 + z^2)^{-3/2} ((a + u) \, du + z\,dz)$. At this point Laplace used approximation A again along with a second approximation, which I shall for convenience call "approximation B." The latter amounts to the equations

$$((a + u)^2 + z^2)^{-3/2} = (a + u)^{-3} = a^{-3} - 3a^{-4}u,$$

with the understanding that whenever any computation is made, all quadratic terms in u and z are to be dropped.

Finally, since the centripetal force needed to keep a point mass in circular orbit at a given angular velocity is directly proportional to the radius of the orbit and directed toward the center of the orbit, this force is $-(a + u)g \, du$ for some constant g. As stated above, the difference between the two gravitational forces and the centripetal force must be normal to the surface of the ring, i.e., parallel to the line with direction cosines proportional to $\partial\phi/\partial u$ and $\partial\phi/\partial z$. In Laplace's language (since these lines normal to the curve determine a differential equation satisfied by the curve) the differential equation of the generating curve has to be obtained by setting these forces equal to zero. Thus ϕ has to satisfy the differential equation

$$0 = \left(\left(\frac{S}{a^2} - ag\right) + \left(\frac{4\pi}{\lambda + 1} - \frac{2S}{a^3} - g\right)u\right)du + \left(\frac{4\pi\lambda}{\lambda + 1} + \frac{S}{a^3}\right)zdz.$$

But since the form of ϕ has already been chosen, the differential equation of

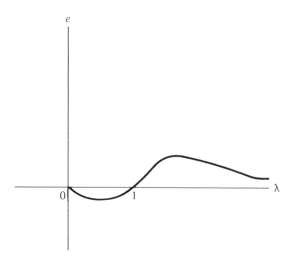

Figure 4-2

the generating curve is easily calculated as $0 = u\,du + \lambda^2 z\,dz$. In order for these two equations to be the same, the coefficients of du and dz in them must be proportional, so that we must have $S/a^2 = ag$ and

$$\lambda^2 = \frac{4\pi\lambda/(\lambda + 1) + S/a^3}{4\pi/(\lambda + 1) - 2S/a^3 - g} = \frac{4\pi\lambda/(\lambda + 1) + S/a^3}{4\pi/(\lambda + 1) - 3S/a^3}.$$

The quantity g is determined by the angular velocity of the rotating ring. The shape of the ellipse is determined by the number λ, which is the ratio of its horizontal and vertical axes. If we simplify things by introducing the convenient number $e = S/4\pi a^3$, the equation for λ becomes

$$e = \frac{\lambda(\lambda - 1)}{(3\lambda^2 + 1)(\lambda + 1)}.$$

The graph of e as a function of λ is shown below in Figure 4-2.

From this work Laplace was able to deduce some very interesting facts about Saturn. For one thing, since by its definition e must be positive, it follows that $\lambda > 1$, i.e., the ring is broader than it is high, a conclusion amply verified by observation. Second, if ρ is the mean density of Saturn, and R its radius, then $S = (4/3)\pi\rho R^3$, and since $S = 4\pi a^3 e$, we get $\rho = 3e(a/R)^3$. The graph of e versus λ shows that e has an absolute maximum value for $\lambda > 1$, and computation shows that $e < 0.0543026$. Observation makes it appear that a/R is about 2 (I am talking about the observations available to Laplace). It therefore follows that ρ is less than $24(0.0543026)$, which is about 1.3. Though Laplace did not say so, it is clear from observation that λ is very large, i.e., the rings are very thin. It therefore follows that ρ is much less than the estimate given by Laplace. This fact also is confirmed by observation: Saturn has a very low density. As a proving ground for Newtonian mechanics the rings of Saturn have thus shown themselves quite useful, linking two apparently unrelated facts, the thinness of the rings and the low density of the planet.

4.6. Extensions

Laplace went on to remark that his argument applies even if the ring is not assumed to be a surface of revolution, i.e., even when the shape of a longitudinal cross section varies as the sectioning plane rotates. This observation is important, since Laplace showed that a uniform homogeneous ring could not be in stable orbit. Since this part of the work does not relate directly to Kovalevskaya's work and is in fact a different topic entirely, I shall not give Laplace's argument, but rather show quickly by an informal argument of my own why such a ring cannot be in stable orbit. (The only reason this subject is relevant to the work of the present chapter is that stability questions cast doubt on the physical assumptions used by Laplace as a basis for his argument; consequently Kovalevskaya's work was not carried out in detail.)

If a body of constant linear density occupying a circle is attracted by a point mass inside the circle but off-center, the center of the circle will move *away* from the attracting mass. In fact if the mass is at $(x, 0)$ and the equation of the circle is $u = \cos \alpha$, $v = \sin \alpha$, $0 \le \alpha \le 2\pi$, then the vertical component of the force on the circle is zero, and the horizontal component is a positive multiple of

$$F(x) = \int_0^{2\pi} (x - \cos \alpha)((x - \cos \alpha)^2 + \sin^2\alpha)^{-3/2} \, d\alpha.$$

Computation shows that $F(0) = 0$ and $F'(0) = -\pi$. It follows that $F(x) < 0$ for small positive x, so that if the center of the circle ever departs from the attracting mass, the attraction of the mass will reinforce, rather than counteract, the perturbation. It can be shown that the circle will move until it touches the attracting point. Since a uniform homogeneous ring is a union of such circles, such a ring would have to crash into the attracting planet if it ever moved slightly off-center. Therefore it could not be in stable orbit. If the ring is in stable orbit and either solid or liquid, its cross section must have variable curvature.

Of the two questions discussed by Laplace, shape and stability, the latter attracted much more attention than the former. James Clerk Maxwell won the Adams prize at Cambridge in 1857 for a thorough study of the stability of the rings (Maxwell 1859). He showed that a solid ring could be stable only if its density were so irregular that it approximated a satellite, and that a liquid ring would be destroyed by resonant waves. It would seem to follow that the rings must consist of discrete particles. Bessel (1807) studied the effect of the gravitational attraction of the rings on the shape of Saturn, but in his study of Saturn's fourth satellite (1812) he assumed that the rings were infinitely thin. Laplace and Kovalevskaya seem to have been the only authors to write about the shape of the rings, though the main problem in the analysis–computing the potential of a ring–was studied by others (cf. Riemann, *Werke*, pp. 431–436).

4.7. Reformulation of the Problem

The motivation for Kovalevskaya's paper was provided by doubt as to the legitimacy of what has been called above approximations A and B. Kovalevskaya said that when a computation is made up to third order (such as approximation B) one should not introduce approximations such as approximation A for which the magnitude of the error is unknown. She said further that even if the assumption of an elliptical cross section does give a solution up to third order, one ought to show that the exact solution can be obtained by a perturbation of the ellipse. She thus set herself the task of finding a theoretically exact solution which can be approximated by an ellipse with known order of error. Kovalevskaya's model for a cross section of the ring was different from that of Laplace, though very like the one used by Riemann (*Werke*, pp. 431–436). It is not clear whether she knew of Riemann's work.

To obtain mathematical conditions which give the shape of the ring, Kovalevskaya used the fact that the potential and kinetic energies of a particle of the ring must have constant sum, that is,

$$V(\rho_1, z_1) + M(\rho_1^2 + z_1^2)^{-1/2} + \tfrac{1}{2}n^2\rho_1^2$$

must be constant. Her notation, which conflicts with that of Laplace, is the following: The cylindrical coordinates of a molecule of the ring are (ρ_1, θ_1, z_1); M is the mass of Saturn, V is the gravitational potential of the ring, n is the angular velocity with which the ring rotates.

Kovalevskaya changed the parameters which describe a cross section of the ring. Her picture was as follows: The distance from Saturn to the midpoint of the horizontal diameter of the generating curve (which Laplace had called a) is the unit of distance, i.e., this distance is taken as 1. The length of half of this diameter (which Laplace had called k) is denoted a. Finally, the

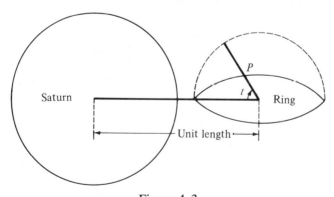

Figure 4-3

equation of the generating curve, which Laplace had taken as $y^2 = \phi^2(x)$, is given parametrically:

$$x = 1 - a \cos t,$$

$$z = a\phi(t),$$

where $0 \le t \le 2\pi$. The geometric significance of this parametrization is shown in Figure 4-3, where P denotes the point (x, z). The radius of the semicircle is a.

The function $\phi(t)$ clearly has period 2π, and both observation and theoretical symmetry amply justify the assumption that $\phi(-t) = -\phi(t)$. Therefore under the reasonable assumption (unstated by Kovalevskaya) that ϕ has a continuous derivative, this function is the sum of a Fourier sine series

$$\phi(t) = \beta \sin t + \beta_1 \sin 2t + \beta \sin 3t + \cdots .$$

The elliptic form of the curve assumed by Laplace is obtained as the special case which results when $\beta_j = 0, j = 1, 2, \ldots$. The only nonzero coefficient (β) then equals the quantity which in Laplace's notation was k/λ.

4.8. Kovalevskaya's Method

At this point it may help to give an overview of Kovalevskaya's method, so as not to get lost in details. Starting from the assumed form of the ring, we can present this program as a four-step process.

(1) Compute the potential of the ring at the point whose cylindrical coordinates are $(\rho_1, \theta_1, z_1) = (1 - a \cos t_1, \theta_1, a\phi(t_1))$. Since this potential is independent of θ, it can be written as a function of t_1. Because of the symmetry of the ring, this potential is an even function of t_1. If its Fourier series expansion is $V = V_0 + V_1 \cos t_1 + V_2 \cos 2t_1 + \cdots$, computing the potential V means expressing the coefficients V_j in terms of the β's. As we saw in the discussion of Laplace's work, the computation of the potential of the ring is the hardest part of the problem.

(2) Similarly, write the other terms in the expression for total energy as Fourier series, i.e.,

$$M(\rho_1^2 + z_1^2)^{-1/2} = m_0 + m_1 \cos t_1 + m_2 \cos 2t_1 + \cdots$$

with an even simpler series for $\frac{1}{2}n^2\rho_1^2$, since $\rho_1^2 = (1 - a \cos t_1)^2$. As in part (1) of the program, this job is considered done when the coefficients in the series are expressed in terms of the undetermined coefficients $\beta, \beta_1, \ldots$.

(3) Write the energy equation $V + M(\rho_1^2 + z_1^2)^{-1/2} + \frac{1}{2}n^2\rho_1^2 = C$ as the system

$$
\begin{aligned}
V_0 + m_0 + \tfrac{1}{2}n^2(1 + \tfrac{1}{2}a^2) - C &= 0, \\
V_1 + m_1 - n^2 a &= 0, \\
V_2 + m_2 + \tfrac{1}{2}n^2 a^2 &= 0, \\
V_j + m_j &= 0, \qquad j = 3, 4, \ldots .
\end{aligned}
\tag{3}
$$

In each of these equations V_j and m_j are presumed to be expressed in terms of the β's. The quantities n and C are theoretically determinable by observation, but the mathematical question being investigated is whether there are any values of n and C which make the motion possible for some shape of cross section. Thus in the theoretical investigation n and C may also be treated as unknowns.

(4) Solve the system (3) for n, C, and the β's. Since each of the equations contains in general an infinite set of unknowns (the β's), this part of the problem leads to new mathematical difficulties. Kovalevskaya suggested in a nonrigorous fashion a method for handling these difficulties which may be the most important achievement in the paper. This point will be taken up after the discussion of the paper.

4.9. Execution of the Program

Let us begin at the beginning, computing the potential of the ring. Kovalevskaya made use of a result of Gauss to which she gave the reference (*Werke* V, p. 286), what is now commonly known as Gauss' theorem (the divergence theorem). The gravitational potential at the point P_1 of a solid body occupying the region B of space is given by the integral $V = -\frac{1}{2}\iint \cos\theta \, d\sigma$, where the integral is taken over the boundary of B, θ is the (variable) angle between the line from P_1 to a (variable) point on the surface of the body and a line normal to the surface at that point, and $d\sigma$ is the element of surface area. Since Kovalevskaya was interested in a ring-shaped body, she wrote $P = (x, y, z)$, $P_1 = (x_1, y_1, z_1)$ and parametrized the surface by two angles t and ψ which range from 0 to 2π, viz.,

$$x = (1 - a \cos t) \cos \psi,$$

$$y = (1 - a \cos t) \sin \psi,$$

$$z = a\phi(t),$$

with similar expressions for the coordinates of P_1 in terms of angles t_1 and ψ_1. Then a straightforward but tedious computation with one change of variable, $\zeta = \frac{1}{2}(\psi - \psi_1)$, gives $V = \int_0^\pi W \, dt$, where

$$W = \int_0^{\pi/2} \frac{C - Aa\phi'(t) \sin^2 \zeta}{(B + A \sin^2 \zeta)^{1/2}} \, d\zeta,$$

$$A = 4(1 - a \cos t)(1 - a \cos t_1),$$

$$B = a^2((\cos t - \cos t_1)^2 + (\phi(t) - \phi(t_1))^2),$$

$$C = 2a^2(1 - a \cos t)((\cos t - \cos t_1)\phi'(t) + (\sin t)(\phi(t) - \phi(t_1))).$$

The integral which defines W can be reduced to complete elliptic integrals with modulus k, where $k^2 = B/(A + B)$, so that the complementary modulus k' satisfies $(k')^2 = A/(A + B)$. When these functions are introduced, the result is

$$W = (A + B)^{-1/2}(C + aB\phi'(t)) \int_1^{k^{-1}} (s^2 - 1)^{-1/2}(1 - k^2 s^2)^{-1/2} \, ds$$

$$- a\phi'(t)(A + B)^{1/2} \int_1^{k^{-1}} k^2 s^2 (s^2 - 1)^{-1/2}(1 - k^2 s^2)^{-1/2} \, ds.$$

The two integrals which occur in these expressions are easily converted into the complete elliptic integrals K' and E' with modulus k' (cf. Section 3.3). (Kovalevskaya did not use the letters K' and E'. She simply wrote down the integrals.) She then applied well-known formulas from elliptic integrals to get

$$K' = \frac{1}{2} K \log \frac{16}{k^2} - \overline{K},$$

$$E' = \frac{1}{2} F \log \frac{16}{k^2} - \overline{F},$$

where Kovalevskaya's K differs from the complete elliptic integral of first kind with modulus k only by the factor $2/\pi$, i.e., it is $2/\pi$ times what is nowadays denoted K. The terms $\overline{K}$ and $\overline{F}$ are defined by the equations. The important fact about them is the fact that they have the series expansions

$$\overline{K} = \frac{1}{2} k^2 + \frac{21}{128} k^4 + \cdots ,$$

$$\overline{F} = -1 + \frac{1}{4} k^2 + \frac{39}{192} k^4 + \cdots .$$

From these formulas Kovalevskaya obtained $W = W_1 \log (16/k^2) + W_2$, where

$$W_1 = \tfrac{1}{2}(A + B)^{-1/2}(C + aB\phi'(t))K - \tfrac{1}{2}a(A + B)^{1/2}\phi'(t)F,$$

$$W_2 = -(A + B)^{-1/2}(C + aB\phi'(t))\overline{K} + a(A + B)^{1/2}\phi'(t)\overline{F}.$$

[The formulas given in Kovalevskaya's collected works (Raboty, p. 144) are incorrect at this point; the square roots in the second terms are placed in a denominator.]

At this point remember that we are trying to express $V = \int_0^{2\pi} W \, dt$ as a Fourier series in t_1. The strategy is therefore to write W as a double Fourier series in t and t_1. The series for V can then be obtained by discarding the terms which explicitly depend on t, since the integral over a complete period of cosine is zero. If the terms which remain are then multiplied by 2π, the result will be the Fourier series of V. The seemingly unmotivated sequence of transformations which Kovalevskaya has introduced is apparently intended to make this computation feasible. The power series for $\overline{K}$ and $\overline{F}$ presumably make the Fourier series of W_1 and W_2 computable. The difficulties then concentrate in the expression $\log(16/k^2)$, since k is small. Considering the value of k^2 given above, the difficulties can be further localized in the last term of the expression

$$W_1 \log \frac{16}{k^2} = W_1 \log 16(A + B) - W_1 \log B.$$

To overcome the difficulties with this term, Kovalevskaya remarked that B is "divisible by $1 - \cos(t - t_1)$," apparently meaning that the quantity B_1 defined by the equation

$$B = a^2(1 - \cos(t - t_1))B_1$$

is an analytic function of t and t_1. The computational difficulties now reside in evaluating the integral of $\log B_1$.

Now, assuming the elliptical form of Laplace is a good approximation to the actual cross section of the ring, all the coefficients in the original Fourier sine series for $\phi(t)$ are very small; consequently B_1 does not differ by very much from its value in the case of elliptical cross section, which happens to be

$$(1 + \beta^2) - (1 - \beta^2)\cos(t + t_1).$$

Kovalevskaya, having obtained this approximation, pushed the difficulties one final step farther away, defining B_2 by the equation

$$B_1 = ((1 + \beta^2) - (1 - \beta^2)\cos(t + t_1))(1 + B_2)$$

and thereby obtaining

$$W_1 \log B = 2W_1 \log a - W_1 \log[(1 + \beta^2) - (1 - \beta^2)\cos(t + t_1)]$$
$$- W_1 \log(1 + B_2) - W_1 \log[1 - \cos(t - t_1)].$$

Theoretically at least, one can now integrate termwise to obtain the desired Fourier series. Because B_2 is small, the only term of this last expression requiring any justification for this operation is the final one. But it involves known elementary functions. Indeed, substituting $\cos(t - t_1)$ in the MacLaurin series of $\log(1 - x)$ yields a series which can be estimated using Stirling's formula. (Kovalevskaya, however, did not give this justification.)

At this point, in principle, step (1) of the program has been carried out. As has been noted repeatedly, this step is the difficult one. Steps (2) and (3) present no particular difficulties. The remaining difficulties occur in step (4), actually solving for the β's. To carry out this procedure Kovalevskaya gave an algorithm which she thought could be used, as follows: (a) Fix a positive integer μ; (b) assume $\beta_j = 0$ for $j > \mu$; then solve the first $\mu + 3$ equations in the system for $n, C, \beta, \beta_1, \ldots, \beta_\mu$, denoting these solutions $n^{(\mu)}, C^{(\mu)}$, etc.; (c) prove that each of these quantities tends to a limit as μ tends to infinity and that the limiting solution satisfies system (3).

The suggested procedure is very natural and in fact is one which is sometimes used today for handling such systems. But for the lack of any *theorems* giving conditions under which the technique will work, Kovalevskaya's paper might have been one of the outstanding early papers in this area. As it is, nothing is proved, and the results are merely suggested. However, one cannot blame Kovalevskaya for not providing the rigor. To do so she would have had to create much of the work of Gram and Schmidt on vector space theory 20 years before it was actually done. Some of the ideas which she suggested without proof were subsequently stated and rigorously developed by Hammerstein (1930). (I am indebted to my colleague J. W. Burgmeier for this reference.) It is fairly clear that enormous practical difficulties arise in the computation of V and in the solution of system (3).

Kovalevskaya pretty much confined her explicit calculations to finding $\beta^{(1)}$ and $\beta_1^{(1)}$, i.e., to the case when $\beta_j = 0$ for $j > 1$. Setting $\beta_1^{(1)} = \beta_1 = a\gamma$, she obtained the following equations for γ in terms of β and for β in terms of n and γ:

$$\gamma = \frac{1}{4} \frac{3M(2 + 3\beta^2)(1 + \beta)^2 - 4\pi\beta(1 - \beta)(1 + 3\beta)}{4 - 12\beta - 12\beta^2 - 12\beta^3 - 3M\beta(1 + \beta)^3}(1 + \beta)$$

and

$$n^2 = M + a^2 N,$$

where

$$N = \left(\frac{-11 + 9\beta - \beta^2 + 3\beta^3}{4(1 + \beta)}\right) + \log \frac{256}{a^2(1 + \beta)^2} + M\left(\frac{6 - 3\beta^2}{8}\right)$$

$$+ \left(2\pi \frac{1 + \beta - \beta^2}{(1 + \beta)^2} - \frac{1}{2}M\beta\right)\gamma.$$

From a practical point of view the most important information to be gained is the sign of β_1, which is the same as the sign of γ. When γ is positive, the oval-shaped curve which generates the ring points away from the planet; when γ is negative, the curve points toward the planet. From the graph of e as a function of λ (given above) one sees that in general small values of e correspond to two distinct values of λ. Kovalevskaya remarked that these two values of λ give distinct signs to β_1, provided M is not too large. In his exposition of this work the astronomer Tisserand (1892, p. 153) carried the reasoning one step further and showed that the border between these two cases occurs when $\beta = 0.24$.

4.10. Significance of the Paper

Work of the sort just discussed may have value of different kinds. It may, for instance, give useful or otherwise pleasing information about the natural phenomenon under investigation. Alternatively, it may serve as a proving ground for physical theories. Yet again, it may contain mathematical techniques which can be applied to other problems. By way of illustration Laplace's work discussed above satisfies the first criterion in that it gives a theoretical upper bound to the density of Saturn. It satisfies the second, since Laplace was able to establish a theoretical equilibrium for a ring of elliptical cross section using reasonable approximations. Finally, the series approximation mentioned, but not discussed, above was a technique of wide applicability (though probably not new when used by Laplace).

Except for the theoretical relation between the value of β and the orientation of the oval cross section deduced by Tisserand, it cannot be said that Kovalevskaya's work satisfies the first criterion. It does contribute according to the second criterion, since it shows how an equilibrium might be theoretically attained without using approximations. Its main value, however, seems to fall under the third criterion.

It was astute, though probably not original, of Kovalevskaya to raise the issues of stability and error analysis which are naturally associated with the approximations used by Laplace. Her alternate method of computing the potential of the ring, breaking it up into terms and applying the series expansions for the complete elliptic integrals, is of use in other contexts. For example, Poincaré (1885) independently used such a technique in a paper on hydrodynamics, and, as already mentioned, Hammerstein (1930) developed her suggestion for solving an infinite system of equations into a rigorous theory.

It may be of value to see why Kovalevskaya did not herself carry out any detailed computations. Even in the most drastically simplified special case where $\beta_j = 0$ for $j > 1$, $\beta_1 = a\gamma$, and all terms which are of third order are discarded, the quantity B_2 is given by the expression

$$a^2\gamma^2 + a\beta\gamma(\cos t + \cos t_1) - 2a\beta\gamma(\sin t \sin 2t_1 + \sin 2t \sin t_1)$$

$$\times \frac{-2a^2\gamma^2 \sin 2t \sin 2t_1 - a\beta\gamma(\cos 3t + \cos 3t_1) - \frac{1}{2}a^2\gamma^2(\cos 4t + \cos 4t_1)}{1 + \beta^2 - 2 \cos t \cos t_1 - 2\beta^2 \sin t \sin t_1 + \frac{1}{2}(1 - \beta^2)(\cos 2t + \cos 2t_1)} \cdot$$

This Brobdingnagian expression is only one of about a dozen similar expressions which have to be calculated in the very simplified special case, and not even the most complicated of those. Apparently no one except Kovalevskaya and Tisserand has ever attempted to do the general case theoretically or any other special case explicitly.

One reason for the neglect of this subject is that the shape of the rings is not of great theoretical interest. Another possible reason was given by Kovalevskaya herself, in the middle of her paper:

> I have carried out completely this–still very long–computation for $\mu = 1$, mostly in order to test the extent to which Laplace's result is approximately true, and to obtain, at least in the large, the deviation of the cross-section from elliptical form. However, I have been deterred from a yet more precise determination of the cross-section of the ring not only by the difficulty of the calculation, but also by the fact that, due to Maxwell's research (On the stability of the motion of Saturn's rings, Cambridge 1859) it has become doubtful whether Laplace's view of the structure of the rings of Saturn is acceptable.

Coming as it does in the middle of many pages of intricate formulas, this statement has a very ironic sound, as if she were saying, "Anything not worth doing is not worth doing well." Nevertheless, although her techniques are applied to a problem of small intrinsic interest, they are still of interest for their own sake.

It is often stated that Kovalevskaya "discovered" that the true shape of the cross section was oval, rather than the ellipse supposedly "discovered" by Laplace. In fact neither of these findings was a discovery. The elliptical form of Laplace was an assumption. The oval form of Kovalevskaya was not a fact at all, but an *arti*fact; that is, it was the result of her decision to use Fourier series to represent the cross section. For the equations

$$x = 1 - a \cos t; \qquad z = a\beta \sin t,$$

and

$$x = 1 - a \cos t; \qquad z = a(\beta \sin t + \beta_1 \sin 2t)$$

represent an ellipse and an oval, respectively, as any freshman calculus student should be able to prove. If Kovalevskaya had wanted that result, she need not have written more than six lines. If she had chosen some other parametric representation for the cross section, the second approximation might not have been oval.

One final comment seems justified before leaving this topic. The paper just discussed seems to be the only one of Kovalevskaya's papers which merits the designation "applied mathematics" in the strict sense. It is the only paper in which she sacrificed rigor for the sake of obtaining a realistic model. This point will be taken up again in Chapter 9.

4.11. Personal Notes

This paper seems to have been included in Kovalevskaya's dissertation at the last moment, at least judging from the trouble she had getting it together. As Weierstrass wrote to Fuchs in requesting a degree for Kovalevskaya on 19 July 1874 (Mittag–Leffler 1923f, pp. 253–254):

> . . . Your last letter delayed somewhat the dispatch of the dissertation since it seemed to imply that the dissertation on the form of Saturn's rings should be submitted together with the dissertation "On the theory of partial differential equations." In the former extensive calculations are omitted whose detailed exposition is superfluous for the stated purpose and it is edited in such a way that Weber will easily be able to render his decision, which will help move the whole matter forward. . . .

Then, two days later,

> I have become a genuine tormentor for you, but what can I do? My pupil just appeared, very upset and blaming herself for not including one page of her work

on Saturn's rings (with which she really did struggle mightily) and doesn't
know what to do. Therefore I must make friend Fuchs my intermediary and ask
him to take care to include this sheet with the others. . . .

In this last quotation Weierstrass is making a pun. The sentence in German
reads: Soeben kommt meine Schülerin in großer Aufregung zu mir, sich
anklagend, daß sie einen Bogen ihrer Arbeit über den Ring–mit dem sie in der
Tat hat sehr ringen müssen–nicht mit eingepackt habe. . . .

    Kovalevskaya herself seems to have thought seldom about this paper.
She decided to publish it only after she became a dozent at the University of
Stockholm, as she said in a letter to Mittag–Leffler quoted in Section 5.8,
because it was a good time to publish as much as possible. To get it published
she relied on the influence of her friend Hugo Gyldén, who inserted it in the
Astronomische Nachrichten.

PART II

MATURE LIFE

CHAPTER 5

Biography: 1875–1891

5.1. Return to Russia

Sonya, Vladimir, and Julia Lermontova shared a moment of triumph in the fall of 1874. All three had been awarded doctorates, Lermontova in chemistry from Göttingen and Vladimir in geology from Jena (in 1872). Vladimir, however, had failed the qualifying examination necessary to procure a teaching post in Russia (cf. Koblitz 1983, p. 118). Sonya was feted on her name-day, 17 September on the Russian calendar; her old tutor Malevich toasted "the health of the first Russian woman scholar." Lermontova was given a warm reception in Petersburg by the chemist Mendeleev. Sonya and Vladimir had reconciled during 1873 and from then on began to make plans for a future together. At some time during the next few years they finally consummated their marriage after seven years of a platonic relationship.

According to Koblitz, Vladimir's letters from this period show a very unrealistic expectation of offers of employment from many sources. Neither Sonya nor Vladimir was ever very far from the radical circles of the time, so that their youthful hopes for a satisfying personal life were combined with a sense of participation in projects of reform of the greatest importance for the future of humanity. Their return to Russia came about the time of the great Populist movement in Russia. Some 3000 upper-class young people, with much idealism but little common sense, took menial jobs in the villages of Russia, some to perform social service, others to spread revolutionary ideas. Of course, the dress and manners of many of them were ludicrously inappropriate for gaining the confidence of the people. It was therefore very easy for the government to eliminate the radicals in the movement.

One of Kovalevskaya's plans had been to consolidate her rather haphazard education. She had confided this plan to Weierstrass, who wrote to her on New Year's Day 1875 (Kochina 1973, p. 58):

> I am in complete agreement with your plan to spend this winter filling in the gaps in your knowledge of the most elementary parts of mathematics, specifically analytic mechanics and mathematical physics. But, along with the British, study also Poisson and Cauchy and the works of the elder Neumann on electrodynamics. The book of the younger Neumann (the son) on this subject is written rather ponderously, but nevertheless contains much that is good and useful. If you have received the extremely thick book of Hamilton on Quaternions by now, I think its mere appearance will have frightened you away from any thorough study of it. On the other hand, in my opinion such work would be a useless waste of time, and I have always pitied those poor students for whom, while Hamilton was alive, quaternions were an obligatory subject.

You need to master many more necessary and useful things than a special method which perhaps is completely usable, but in no case necessary, for the solution of certain problems and whose algebraic foundations are of a very trivial nature. I am sure that this food is not to your taste. If time permits you might also look at the *Comptes Rendus* of the French Academy, paying particular attention to the works of Saint-Venant on the theory of elasticity. To be sure, you will be impressed throughout (i.e. reading the authors I have named) by the lack of rigor in the proofs. But do not be discouraged by this; after all the important thing is for you to get some idea of what has been done up to now in mathematical physics and of unsolved problems. In doing this you can work on some easy problems to practice exposition, where, as I have often told you, the elegant working out of details should be considered an essential thing.

The last sentence is an allusion to an earlier part of the same letter, where Weierstrass had written:

I do not consider myself a scientific pedant and I do not claim that there is only one True Church in mathematics. However, what I do demand of a scientific work is unity of method, the sequential following of a definite plan, and the appropriate working out of details, and that it should bear the stamp of independent investigation.

It is worth emphasizing once again that these words reveal some basic principles of Weierstrass' approach to mathematical exposition, the careful attention to details. In particular, if a pure existence proof can be replaced by a constructive proof, then the constructive proof is to be preferred: a bird in the hand is worth two in the bush. These principles help explain Kovalevskaya's extraordinary efforts to obtain explicit formulas in her prize-winning work on the Euler equations for a rotating rigid body.

For Sonya and Vladimir disillusionment in Russia was not long in coming. To teach above the elementary level in Russia, even with a German doctorate, one had to pass the magisterial examination, which, as just mentioned, Vladimir had failed in 1873. Sonya was not even allowed to take the examination. Vladimir finally passed the examination in March 1875, but still received no offer of a job which attracted him (Koblitz 1983, p. 128). Only Julia Lermontova, after a long time, managed to find a position in a laboratory.

The response that Sonya and Vladimir made to this frustration was to speculate in real estate in the hope of becoming independently wealthy and thus free to pursue their scholarly interests. This project became a veritable mania and eventually involved Sonya's uncle Fyodor Fyodorovich Schubert, Vladimir's brother Alexander, and even Julia Lermontova. The death of General Korvin-Krukovsky in 1875 brought Sonya an inheritance of 50,000 rubles, less 20,000 previously loaned to cover the debts of Vladimir's publishing business. At this point it appeared that they had procured enough capital to provide a secure income if carefully invested.

While engaged in this business activity the couple settled into a rather active social life in Petersburg, frequently meeting some of the most famous scholars and writers in Russia. Mendeleev has already been mentioned as one of their acquaintances. They also had visits with the mathematicians P. L. Chebyshev and G. Mittag-Leffler (from Helsingfors) and with the writers Turgenev, Dostoevsky, Nekrasov, and Saltykov-Shchedrin. Although one cannot often assume that a work of fiction is strictly autobiographical, it is difficult not to believe that in the following passage from her 1892 novel *Nigilistka*–The (fem.) Nihilist–Kovalevskaya was describing her own experience of a decade-and-a-half earlier. (Two explanatory notes are included after the passage.):

I was twenty-two years old when I moved to Petersburg.[1] About three months earlier I had graduated from a foreign university and returned to Russia with a doctor's diploma in my pocket. After five years of lonely, almost reclusive, life in a small university town Petersburg life immediately seized me and seemingly intoxicated me. Forgetting for the time being those concepts about analytic functions, about space, about four dimensions, which had until recently filled my whole internal world, I was plunging heart and soul into new interests, meeting people right and left, trying to penetrate into the most diverse circles and hearkening with eager curiosity to all the happenings of that complicated bustle, so empty in reality and so engrossing at first glance, known as Petersburg life. Everything interested and delighted me now. Everything amused me—theaters, charity balls, and literary circles with their endless futile arguments on all variety of abstract themes. The regular visitors to these circles had already had time to get fed up with these arguments, but for me they still had all the charm of novelty. I abandoned myself to them with all the ardor possible to a Russian who is garrulous by nature and has just spent five years in Germanity [*v nemetchinye*], in the exclusive society of two or three specialists, each occupied with his own narrow affair and not understanding how anyone could spend valuable time on the frivolous grooming of the tongue. The pleasure that I myself experienced from socializing with other people spread to those around me. Being myself amused, I injected new liveliness and life into the circle I moved in. The reputation of learned woman surrounded me with a certain aura; acquaintances all expected something of me; two or three newspapers had already managed to trumpet me about; and this, for me, completely new role of eminent woman, although it embarrassed me somewhat, nevertheless amused me very much at first. In short, I was in the most benign frame of mind; I was, so to speak, passing through my honeymoon of fame and at this period of my life, would gladly have exclaimed, "All is arranged for the best in this best of all worlds!"

In October 1878 Sonya gave birth to a daughter, Sophia Vladimirovna Kovalevskaya, nicknamed Fufa by her parents. Fufa's first year of life was not filled with good omens. Her grandmother Elizaveta Korvin-Krukovskaya died in February 1879 and soon afterward her parents investments failed. The

---

[1]Kovalevskaya was actually 24 when she settled in Petersburg. Even when she was *not* writing fiction, she frequently set her age back a few years.

financial disaster affected Vladimir very deeply, and this heavy blow was followed by one even more severe for him. He found that one of his radical acquaintances (a paranoid lot to begin with) suspected him of being a tsarist spy (Koblitz 1983, p. 139). With no secure position in the world of business, academia, or radical politics, this very high-strung young man lost all interest in work and people and withdrew from society.

Sonya, on the other hand, responded more positively to the crisis. During a difficult pregnancy and a long puerperal illness she had had time to reflect on the direction of her life. Evidently she had already decided to make some changes even before the crisis, for she had resumed writing to Weierstrass in the fall of 1878 after a two-year hiatus. No doubt she felt keenly the waste of the education which had cost her so much toil and struggle. The birth of her daughter must have given extra urgency to the need to demonstrate the potential a woman possessed, even as the new responsibilities of motherhood made it more difficult to do so. During the financially troubled year of 1879, with an infant daughter, she could hardly have found much time for mathematics. Nevertheless she talked with Chebyshev and at his suggestion gave a paper at the Sixth Congress of Mathematicians and Physicians held in January 1880 (late December 1879 on the Russian calendar). This small exposure to the society of mathematicians turned out to have great importance for her future; for among her audience was one slight acquaintance of hers with an unusual zeal for organizing mathematical activity. This man was the Swedish mathematician Gösta Mittag-Leffler.

The fact that Kovalevskaya was able eventually to pursue a mathematical career when earlier women such as Sophie Germain, who were also of great talent, had been confined to private study and correspondence, can be explained by four factors:

(1)   Kovalevskaya's unusual combination of talent and determination,
(2)   the social movements of the time, which had prepared the way for the acceptance of professional women,
(3)   the influence of Weierstrass,
(4)   the extraordinary efforts of Mittag-Leffler.

Certainly the first two factors were essential. Whether they would have been sufficient of themselves to procure a career for Kovalevskaya is difficult to say. Whatever the case, Kovalevskaya was indeed fortunate to have the support of Weierstrass, whose enormous prestige secured a fair hearing for her dissertations and caused her to be listened to when she spoke about mathematics. Note, however, that in requesting a degree for her Weierstrass insisted in the strongest possible terms that his own involvement be discounted (Mittag-Leffler 1923f, p. 249). Although many countries seemed about ready to admit women to their university faculties, the honor of being the first fell to Sweden, and that largely because of the activity of one young man in a completely new university.

5.2. Mittag-Leffler

Kovalevskaya's professional elder brother was born Gösta Leffler in Stockholm in 1846. The following sketch of his life is based on Weil (1982), Kochina (1981), Grattan-Guinness (1971), and Hille (1962). His father, Johann Olaf Leffler was an elementary school teacher, later (1867–1870) a member of the Riksdag. His mother, Gustava Wilhelmina Leffler, née Mittag, was a highly respected member of Stockholm society. There were three other children in the family. Gösta's younger brother Fritz became a poet and professor of Nordic languages at the University of Uppsala. His sister Anne-Charlotte became a well-known writer (collaborating on several works with Kovalevskaya) and, upon her second marriage, Duchess of Cajanello. The younger child, Artur, became an engineer.

At some point Gösta decided to adopt his mother's maiden name and was thereafter known as Mittag-Leffler. In 1872 he received the doctor's degree from the University of Uppsala for his dissertation, "On the division of the roots of a synectic function of one variable." That same year he became a docent at the University of Uppsala. From 1873 to 1876 he traveled to Paris, Göttingen, and Berlin on a stipend. He was offered a position at Berlin in 1876, but refused it and instead accepted a chair at the University of Helsingfors (now Helsinki).

In 1876 Finland, which had previously been under Sweden's influence, was a dependency of Russia known as the Grand Duchy of Finland. The constitution which had been granted by Tsar Alexander I more than 50 years earlier specifically left the law courts and coinage to the Finns and made the government bureaucracy "non-Russian," which meant in practice, largely Swedish. Nationalism ran high among the Finns, and Mittag-Leffler was soon embroiled in controversy, exacerbated by his efforts to obtain an appointment for Kovalevskaya. Weierstrass described the situation as he saw it in a letter to Kovalevskaya of 15 August 1878: (Kochina 1973, pp. 80–81).

> . . . Mittag-Leffler has become one of my favorite students; he possesses, besides a thorough knowledge, an extraordinary ability to learn and a mind directed to the ideal. I am sure that contact with him would have a stimulating effect on you.
>   The position he occupies in Helsingfors is very unpleasant. There, more than anywhere else, an attempt has been made to create–one would say–a national Finnish mathematics. Articles have appeared in the local newspapers every semester since Leffler has been there attacking Weierstrassian mathematics, for Leffler has been indiscreet in mentioning my name in his lectures and articles more often than necessary. . . .

Thus, when Mittag-Leffler began his efforts to obtain a position for Kovalevskaya (in 1880) he was already in a precarious position himself. He persisted in his efforts, and eventually made his own position completely untenable. Ironically, the efforts came to grief, not because Kovalevskaya was a woman, but because she was a Russian. As he wrote to her on 23 March 1881:

. . . Nearly everyone here has very liberal opinions on the women's question and one could hardly find a single professor who would be against hiring a woman here as Privat-docent provided she was a Finn. As for you personally, all my friends at the university are so convinced of your exceptional talent that there is no doubt you would be hired here if you were Finnish, or indeed belonged to any country except Russia. Just now, however, there are quite different problems. [Tsar Alexander II had been assassinated two weeks earlier.] As you know, Finland is not a province of Russia but a separate country joined to Russia and whose Grand Duke is the emperor of Russia. Our university also has a very good independent position and there is not the least trace here of the leftist movements found in Russian universities. It is thus clear that the authorities will not allow this position to be compromised under any circumstances. Once you were here it is extremely probable that you would be followed by many Russian women students, and it is impossible to ensure that among these students there would not be one who happened to belong to the revolutionary party. But one such incident could pose very severe difficulties for us, which we would never wish for our own compatriots. This is the only real reason why it will not be possible for many years to hire a Russian woman as professor at the University of Helsingfors.

But Madame Pokroffsky has told me that you would be willing to accept an official position at any university and consequently I take the liberty of asking if you would authorize me to try to procure such a position for you. I am thinking particularly of the new university which is to be founded in Stockholm. I have been informed "sub secreto" that I will be appointed professor of mathematics there–I immediately beg you to keep this confidence strictly to yourself–and I have no doubt that some day it will be easy for me to arrange an appointment for you, initially as "Privat-docent." The Swedish language will not cause you any great difficulty and the capital of Sweden is one of the most beautiful cities in Europe, where you will also find a very large circle of scholars of great merit. . . The celebrated astronomer Mr. Gyldén is a good mathematician. . . .

This letter reveals something of the enthusiasm with which Mittag-Leffler wove his plans and which made him into one of the leading organizers of mathematical activity in the late nineteenth and early twentieth centuries. When planning a project, he does not seem to have taken obstacles very seriously. As the letter also shows, Mittag-Leffler's career in Helsingfors was coming to an end, and he was planning to return to his native city. Despite the turbulence of his career in Helsingfors, Mittag-Leffler did some good work while he was there. The famous theorem of complex analysis which bears his name was published while he was there (Mittag-Leffler 1876; cf. Frostman 1966). This theorem received the extreme compliment of being simplified and recast by Weierstrass (1880 = 1886, pp. 57–66 = *Werke* II, pp. 189–199). No doubt this theorem provided opportunities for Mittag-Leffler to make himself unpopular by citing Weierstrass.

In 1881 Mittag-Leffler resigned his position in Helsingfors and joined the newly founded Stockholm Högskola, which had been founded in reaction to the conservative character of the older Swedish universities of Uppsala and Lund. As professor ordinarius in Stockholm, Mittag-Leffler began a 30-year career of vigorous mathematical activity. In 1882 he founded the *Acta Mathematica,* which a century later is still one of the world's leading mathematical journals. Through his influence in Stockholm he persuaded King Oscar II to endow prize competitions and honor various dintinguished mathematicians all over Europe. Hermite, Bertrand, Weierstrass, and Poincaré were among those honored by the King. No doubt Mittag-Leffler's influence was somewhat increased in 1882 by his marrying Signe Lindfors, who was heir to a large fortune. (It was her dowry which enabled him to found the *Acta*.) Except for one brief occasion when his in-laws attempted to settle Signe's share of the fortune on her brother, Mittag-Leffler was financially independent from the time of his marriage until World War I. He used his wife's fortune to collect the best private library of mathematics in the world and to build an elegant villa in Djursholm, north of Stockholm. This villa, now the Institut Mittag-Leffler, is an active center of current research in mathematics and houses many manuscripts of importance in the history of nineteenth century mathematics which were described by Grattan-Guinness (1971). From this center Mittag-Leffler conducted his multifaceted activity, organizing conferences and competitions and carrying on an extensive correspondence (over 7000 letters, now all neatly arranged) with nearly every mathematician of note in the world.

The late Einar Hille, who met Mittag-Leffler during the latter's semiretirement in 1912, described him as a very energetic and sympathetic, though somewhat dogmatic old man (Hille 1962). Long-time residents of Djursholm have similar recollections (Erikson 1981). They say that he was an early cyclist who, in his cycling as in his lecturing, ran heedlessly over anything that was in his way. One resident recalls that he insisted on having the same chair in every meeting and that he used to spit on the ceiling. These are not the diplomatic qualities one would expect of someone who spent a great amount of time organizing people, but Mittag-Leffler was a man of integrity, who was intensely loyal to his friends and devoted to knowledge.

Although he officially retired in 1911, Mittag-Leffler continued to serve as editor of the *Acta Mathematica* for the remaining 15 years of his life. In World War I he lost a considerable fortune and in order to retain possession of his house had to turn it into a national trust; hence the Institute Mittag-Leffler came into being. He summarized his career in a "swan song" at the International Congress of Mathematicians in 1925 in Copenhagen. He died in 1927.

5.3.   The Year 1880

Vladimir Kovalevsky, as previously mentioned, was pathetically unable to cope with the financial disaster which culminated in the sale of most of the family possessions. Sonya, who was perhaps more used to dealing with this kind of hardship, was more rational. Probably at her instigation the couple moved to Moscow, where there might be some hope for Vladimir to obtain an academic position and forget his financial troubles. His brother Alexander, who was an eminent zoologist, attempted to procure an appointment for him at the University of Moscow. While these efforts were proceeding, Sonya and Vladimir were once again working on other projects. Sonya was applying to take the magisterial examination herself so that she would be formally qualified to teach at Russian institutions of higher learning. She withdrew her application temporarily when she was told that it would jeopardize Vladimir's prospects. Regrettably, Vladimir had been lured into another business venture, this time with a pair of wily oil magnates, the Ragozin brothers. He was unfortunately too gullible to see that his position combined a maximum of responsibility with a minimum of authority. While his appointment in Moscow was pending, he traveled frequently to promote the interests of the company. All this travel must have put a strain on the marriage; quarrels with Sonya became more and more frequent. When Vladimir's appointment was virtually certain, Sonya resubmitted her application for the magisterial examination and studied all summer to prepare for it. In the end, however, she was refused permission even to take the examination. In the archives of the Soviet Academy of Sciences there are records showing that Kovalevskaya had previously applied to take this examination in March of 1875 and that the request was approved by the physico-mathematical faculty on 19 May 1875 (Kochina 1981, p. 288). Kochina believes that the action of the faculty was vetoed by a higher official, and indeed this seems the likeliest explanation.

Although Vladimir had been appointed to a rather menial post at the University of Moscow, to begin January 1881, his distracted mind and his continuing involvement with the Ragozins jeopardized even that position. Sonya very rationally began arranging an independent life for herself and Fufa. While Vladimir was away on one of his business trips in the fall of 1880, she wrote to Weierstrass, the first letter she had sent to him since 1878, to say that she would like to visit him in Berlin. Although Weierstrass was happy to hear from her, he wrote that he was busy and ill, and so would have very little time to help her. His letter arrived too late, however; Kovalevskaya had already left for Berlin by the time it arrived. Little is known of the results of this trip, except that Sonya's close friendship with Weierstrass was renewed, to last for the rest of her life.

Kovalevskaya returned to Moscow in January 1881 to learn that her husband had not appeared to take up his post at the University. She was further shocked to learn that, by the Ragozins' accounting, he owed them a large sum of money. Vladimir finally arrived near the end of February. However, in March of that year Tsar Alexander II was assassinated. With their radical ties, both Sonya and Vladimir might have found life difficult in Moscow. Whether for that reason or some other, they left the country in the spring. Sonya and Fufa again went to Berlin, but Vladimir returned to Russia to visit his brother Alexander in Odessa. It was at this time that Mittag-Leffler's efforts to obtain an appointment for Kovalevskaya in Helsingfors fell through. To the letter quoted above Kovalevskaya replied (Kochina 1981, p. 11).

> . . . I never seriously believed in your plans for Helsingfors, in spite of wanting very much for them to work out. I also do not intend to place too much hope on Stockholm, although I admit that I would be ecstatic if I were given this opportunity. . . .

5.4. A New Project

About this time, if not earlier, Weierstrass performed a most valuable service by giving Kovalevskaya a mathematical project to work on. In a letter of 6 March, he wrote to her, "I need not say explicitly that I am eager to find out how far you have progressed in your work–I mean the λ-work." The letter λ had been used by Lamé to denote the parameter which defines the time at which a wave surface reaches a certain point, and the project Kovalevskaya eventually published was an attempt to solve a differential equation using the wave surface. Years earlier Weierstrass had discovered an extremely clever method of constructing general solutions of a large family of linear partial differential equations, more specifically of constructing what we now call the Green's function for certain initial-value problems. Among the equations he solved by this method were those which we would nowadays write in the following forms (for an historical discussion, see Chapter 6):

$$\frac{\partial^2 u}{\partial t^2} = \operatorname{div}(A \operatorname{grad} u), \ u(\mathbf{x};0) = f(\mathbf{x}), \ D_t u(\mathbf{x};0)$$

$$= g(\mathbf{x}); \text{ a positive-definite}$$

and

$$\frac{\partial^2 \mathbf{u}}{\partial t^2} + a^2 \operatorname{curl}(\operatorname{curl} \mathbf{u}) - b^2 \operatorname{grad}(\operatorname{div} \mathbf{u}) = \mathbf{P}; \qquad \mathbf{u}(\mathbf{x};0) = \mathbf{f}(\mathbf{x}),$$

$$D_t \mathbf{u}(\mathbf{s};0) = \mathbf{g}(\mathbf{x}).$$

Weierstrass had never published this method, and he thought that if Kovalevskaya wrote a paper which used the method to solve a significant problem in mathematical physics, she would thereby gain reentry into the mathematical world. The particular problem he had in mind was the set of equations which Lamé had derived to describe the displacement of a particle in an elastic medium. The problem was of significance at the time because one of the competing theories of light propagation was that light is a disturbance in an elastic medium. The equations which Lamé had given would nowdays be written $\operatorname{div} \mathbf{u} = 0$, $\operatorname{curl}(A \operatorname{curl} \mathbf{u}) = -\partial^2 \mathbf{u}/\partial t^2$, where A is a positive-definite operator. Lamé had been able to find planar waves satisfying these equations. His attempts to find solutions of these equations to describe a wave emanating from a point source, however, had run into difficulties which he himself admitted. (In modern language, his solution had a singularity at the source of the radiation.)

Thus in mid-1881 Kovalevskaya settled down to work on a problem which was probably not of her own choosing. Soon after beginning work on this problem she found herself distracted by new insight into another problem of mathematical physics which had interested her many years earlier, the Euler equations which describe the motion of a rotating rigid body. The historical discussion of these equations is given in Chapter 7. Again at this point I content myself with a brief description in modern notation. The equations are

$$I\frac{d\omega}{dt} + \omega \times I\omega = Mg\mathbf{X} \times \gamma; \quad \frac{d\gamma}{dt} + \omega \times \gamma = \mathbf{0},$$

where I is the operator whose matrix is the inertia tensor for the body, ω is the angular velocity of the body, the origin is the fixed point in the body, γ is a vertical vector of unit length, M the mass of the body, $\mathbf{X}$ is the location of the center of gravity of the body, and g is the acceleration of gravity. (A derivation of these equations from Newton's laws of motion is given in Appendix 4.)

These equations form a system of six first-order differential equations of great symmetry. Despite this great symmetry, however, the general solution had not been found. The integrals needed for the solution were at least hyperelliptic integrals, and probably even more complicated (cf. Chapters 3 and 7). The work of Jacobi (1849) had shown that the theory of theta functions might be profitably used to obtain a very explicit solution in a special case ($\mathbf{X} = \mathbf{0}$) and had led Weierstrass to hope for solutions in more general cases using the theory developed for the Jacobi inversion problem. Evidently he had proposed this problem to Kovalevskaya while she was still a student. At that time she had made no real progress; but in the summer of 1881, when she needed to work on her "bread-and-butter" project, new ideas came to her, just interesting enough to set her chasing two rabbits at once. It is not certain just what these ideas were. Her letter of November 1881, which will be quoted below, seems to imply that she saw something new about the use of theta functions, which were Jacobi's method of solving the special case.

Thus in summer 1881 Sonya was living in Berlin with Fufa and Fufa's nanny and trying to get some work done. She did travel briefly to Marienbad to visit Weierstrass and his sisters at their holiday resort. Vladimir was in Russia, getting deeper into trouble. In fact (cf. Koblitz 1983, p. 158) he had even sold the jewelry Sonya had inherited from her mother. Apparently also he had become bitter and cruelly suggested that Sonya was wasting her time on mathematics. She wrote to him (Kochina 1981, p. 97):

You write with perfect justice that no woman has yet achieved anything, but it is precisely because of this that I must . . . put myself in a position where I will be able to show whether I can achieve anything or whether I don't have brains enough for that.

One should not underestimate the possibility that at a time when it was not yet proved that women can do mathematics Vladimir's slur may really have damaged Sonya's self-esteem. The reply just quoted is not exactly brimful of confidence.

Despite her marital difficulties, Sonya wrote Vladimir some affectionate letters in the summer of 1881. In one of these, from Marienbad, (Shtraikh

1951, p. 259) she wrote that she was prepared to return to Russia if he would find suitable lodgings for the family, but that she needed a little more time to finish her work.

At the Institut Mittag-Leffler, I found a letter from her brother-in-law Alexander Kovalevsky dated 16 September (28 September on the western calendar) which internal evidence positively dates to the year 1881. The letter asks her to take Fufa to the south of France (Marseille) to meet him and his family for a reunion. He goes on to say, "and I solemnly assure you that all your rules about Fufa's upbringing will not only be strictly observed, but will not be criticized." Evidently Kovalevskaya had strong views on the proper way to raise a daughter.

Meanwhile Mittag-Leffler was still trying to procure a position for Kovalevskaya in Stockholm. On 15 July he wrote to her that he was "almost certain" that it would soon be possible to offer her a position as privatdocent. Kovalevskaya's research, however, did not progress as rapidly as she had first believed it would. The "month or two" she had told Vladimir would be sufficient was definitely not enough. The cause of the delay, as we now know, what that she was working on two problems at once. She explained her situation in a letter to Mittag-Leffler of 21 November 1881, which contains so much valuable information about her life and work that it seems advisable to print a full translation of the French original. Three explanatory notes follow the letter.

> Berlin
> Schellingstrasse 16
> 21 November 1881

Dear Sir,
I cannot help beginning my letter by expressing my deep and sincere gratitude for the interest you are taking in me and of which I am even more aware than I can tell you. Excuse me for having taken a week to answer your last two letters. Their subject is so serious that a week was not too much time to consider it. I was very happy that your letter reached me in Berlin. Because of that I was able to consult M. Weierstrass directly and be guided by the opinion of our master. No doubt you know as well as I how much gratitude and friendship binds me to M. Weierstrass and how much interest he has always been kind enough to take in all that concerns me. Consequently you will easily be able to believe that in such a serious matter I shall be guided entirely by him. His opinion in this matter is the following: He thinks that the appearance of a woman in the capacity of docent in a university is such an important step, one capable of having such serious consequences for the very cause I am most ardent to serve, that I do not have the right to decide on it before I have shown what I am capable of thorough works of pure science. M. Weierstrass thinks, consequently, that it is absolutely essential that I first finish the investigations which occupy me at present and to which I have already devoted almost a year, and that until I finish them I should not allow myself to be distracted by anything else, nor accept such serious obligations as those which you have kindly proposed for me. I must admit that I find M. Weierstrass' arguments so just that I can only conform entirely to them. You see consequently, dear Sir, that it is unfortunately out of the question that I take up my duties in the course

of this winter. But, I repeat, once my investigations are completed, if you would again take this matter into hand, I shall be very happy.

Now I shall tell you, if you will allow me to do so, about the investigations with which I am occupied. This past autumn I began to work on the integration of the partial differential equations which arise in optics in the problem of refraction of light in a crystalline medium. This work was quite well along when I had the weakness to allow myself to be distracted by another question, which has never stopped rattling around in my head since almost the beginning of my mathematical studies, and in which, for a time, I feared I would see myself surpassed by others. The problem involves solving the general case of rotation of a heavy body about a fixed point by means of Abelian functions. M. Weierstrass had once previously suggested that I work on this problem, but all my attempts at the time were fruitless; and M. Weierstrass' own investigations showed that the differential equations of this problem cannot be satisfied by single-valued (eindeutig) functions of time.[2] This result compelled me to abandon this problem for a while, but since then the beautiful, still-unpublished, research of our master on the stability of the solar system and the analogy with other problems of dynamics have renewed my zeal and given me the hope of satisfying the conditions of this problem by Abelian functions *whose arguments are nonlinear functions of time*. This research seemed so interesting and beautiful to me that I forgot everything else for a while and abandoned myself to it with all the impetuosity of which I am capable. The route I followed consisted in expressing the variables of the problem by theta functions of two variables which for certain values of the constants reduce to the elliptic theta functions which arise in the particular case of Lagrange,[3] then trying to choose them so as to be able to integrate the differential equations between the theta functions and time. The calculations which I got into in this way were so difficult and complicated that I cannot yet say if I will reach the desired end by this route. In any case I hope to know in two or three weeks at the most what I should do about it, and M. Weierstrass is consoling me that even in the worst case I could always reverse the problem and try to find out which forces lead to a rotation whose variables can be expressed by Abelian functions–a meager problem, to be sure, and one far from having the same interest as the one I have set myself, but one which I shall have to settle for if I have bad luck. I shall have to rely on the example of M. Neumann[4] who chose an analogous problem for his doctoral dissertation.

This, dear Sir, is the state of my affairs. I shall ask only that you please consider everything I have just written you as *completely* confidential and keep it strictly to yourself. Being not yet sure of success, I would not like anyone but my friends to talk about it.

In passing through Petersburg I had a chance to see M. Chebyshev and have a long conversation with him. To my great astonishment, and my great

---

[2]Weierstrass' published works do not seem to contain any such result. With Weierstrass such an absence is not surprising, since so much of his work was communicated in his lectures. To show this result, it suffices to find one case in which the solutions are given by hyperelliptic integrals.

[3]The Lagrange case is that of a body with two principal axes of inertia equal and the center of mass on the third principal axis (cf. Chapter 7).

[4]The allusion is to the dissertation of C. Neumann (the younger Neumann referred to in Weierstrass' letter of 1 January 1875 quoted above). It was entitled "De problemate quodam mechanico, quod ad primam integralium ultraellipticorum classem revocatur" and was written in 1856.

satisfaction as well, I found him changed in many respects. He speaks respectfully of the Berlin school, and for you personally, dear Sir, he expresses a very great admiration; he even confided to me that he is going to try to nominate you for the vacancy in the St. Petersburg Academy, but he fears he will encounter many obstacles in his path. I need not tell you that this also *is confidential*. There is here, among M. Weierstrass' *Zuhörer* a young Russian who was especially sent by M. Chebyshev. Isn't *that* a change!

As I intend to write to you again soon, I shall wait until my next letter to report to you M. Weierstrass' opinion on the new memoir of M. Picard, but I warn you in advance that I did not succeed in making him very eloquent on this subject.

While waiting I hope that you will be so kind as to give me some news about yourself soon, and I assure you of the very real and sincere friendship which I have come to feel for you, dear Sir.

Sophie Kowalevsky

How I regret that circumstances and distance do not allow me to hope to see you soon; I would so much enjoy a good long mathematical conversation with you.

I forgot to tell you that I saw Mlle. Pokroffsky in Moscow. I must confess that I found her, as well as her parents, very discouraged. The language difficulties, combined with the difficulties which beset anyone beginning the study of higher analysis after completing the Gymnasium course, had a totally depressing effect on her. After a long conversation with her I thought it more prudent to advise her to spend this winter in Moscow, where she is studying analysis and geometry under the direction of some quite good masters. For next winter she firmly intends to go to Stockholm.

Adieu again,

S.K.

5.5. Crisis

This period of increasing disorganization in Sonya's marriage and professional life reached a crisis in early 1882. She had gone again to Paris to work alone on her mathematical problems, leaving Fufa with her brother-in-law and his family in Odessa. Vladimir came to Paris to confer with her on their financial difficulties. Kovalevskaya's diary from the month of January, which is in the Institut Mittag-Leffler in Djursholm, Sweden, points to the probable outcome of the difficulties. Not without quarrels the two agreed that Vladimir return to Russia and straighten out the finances as well as he could. A meeting of their creditors had been announced for February, and Sonya wrote that, unless the creditors would agree to a postponement, the situation was desperate. After his departure Vladimir delayed sending her a letter, and she began to get worried. When she finally did hear from Vladimir's brother Alexander that Vladimir was in Nice, just as she received a letter from Vladimir saying that he had returned directly to Russia, she was furious. She asked rhetorically how Vladimir could amuse himself in Nice knowing that the wife he had deceived was sitting in an icy room in Paris. These diary entries show a decision to put the past behind her once and for all. She wrote that perhaps the Lord would help her mark a new era in her life from that point. She became determined to finish the mathematical work which she had begun, and which had been languishing for several months. Significantly she entered in her diary on 18 January, "Sent letter to Weierstrass."

Weierstrass' reply came many weeks later. On 11 April he wrote (Kochina 1973, pp. 90–91):

> . . . Your first letter from Paris was long in coming and I confess freely that it would have been very difficult for me to reply to it immediately. From every line of it, and even more from what could be read between the lines, it was abundantly clear that you were gripped by cares and worries which you did not wish to speak of in detail and which threatened to hinder for a long time your ardent desire to abandon yourself calmly to your work. It is not your habit to express yourself frankly to your friends in such situations and you hold that everyone should try to deal independently with whatever he has to bear. I sympathize completely in this respect and therefore I did not venture to ask you for explanations or details. But even so, as your true friend and spiritual father, I could hardly pass over in silence what you communicate in hints and what I have succeeded in understanding. . . .

After this sensitive and perceptive beginning Weierstrass began telling her of his own latest work and inquiring about her work, after which he mentioned a business matter involving a hundred marks which she had sent him. (Where she got the money and why she sent it to Weierstrass is not clear.) He ends with some news from Paris about the work of Poincaré extending work of Fuchs, Klein, and Schwarz (cf. Mittag-Leffler 1923b). He criticized the young Frenchmen for trying too hard to get into the Académie des Sciences and Hermite for encouraging this tendency.

It is difficult to see how any of this could have helped Kovalevskaya or how, in her state, any of it could have interested her. Still Weierstrass knew her very well, and he was a sensitive man. Perhaps he really was doing his best for her. In his next letter (14 June, 1882) he seems more to the point (Kochina 1973, p. 93).

> My Dear Friend,
>
> What you told me in the first part of your long-awaited last letter disturbed me greatly, though it didn't exactly surprise me. In fact I have long suspected the real reason for your long stay in Paris and your absolute silence with respect to me. The few hours in which I had the chance to become acquainted with Herr K[owalewsky] sufficed to give me the conviction that your relationship had an internal rent which threatened to destroy it completely. He has neither interest in nor comprehension of your ideas and aspirations, and you cannot stand the turbulence of his life. Your personalities are too different for you to hope to gain a base, a support for yourself (which is essential for a happy marriage) and for him to find in you the necessary completion of himself. If this were otherwise, I believe, even certain breaches of faith on his part would not prevent a reconciliation in good time.
>
> If I have believed that I should object to your plan to go to Stockholm as privatdozent while he was waiting for a post in Moscow, this was done in the conviction that such a relationship between spouses is unnatural, and I will not even allow it to be said to me that you would have thought of it if you felt yourself inwardly bound to your husband and had loved him as a husband wishes to be loved. I could not have believed him capable of being against this

plan; perhaps that is precisely the reason he has set himself even more against your mathematical efforts. As the matter lies at present, the previous relations between the two of you seem to have become impossible. It would nevertheless be a good thing for you to gain the freedom from disturbance and care necessary for your existence. You must get out of your present solitude as soon as possible, and also have little Sonia [Fufa] with you–caring for her and seeing after her development will occupy you beneficially and make you happy.

In the preceding I have expressed my view without mincing words.

I thank you for the trust you have shown in me, but I know you too well to try to compel you to take any advice, and I know that you are quite content to deal with your own fate. But if you think that my advice or support could be of any use, you know well that you can turn to me without hesitation. . . .

Weierstrass continued to try to interest Kovalevskaya in mathematics during the summer of 1882. On 15 July he wrote to her (Kochina 1973, pp. 100–101):

. . . Lindemann's work on the number π is remarkable for many reasons; the results are correct, but were at first based on an *incorrectly* understood proposition and even now have not been rigorously proved by Lindemann.

Through certain theorems which belong to the circle of ideas developed by Hermite in his treatise on the exponential function but which do not require so much formal apparatus, I have arrived at a completely rigorous and not difficult proof of the general Lindemann theorem, namely the following:

'If $z_0, z_1, \ldots, z_n$ are distinct algebraic numbers and $N_0, N_1, \ldots, N_n$ are arbitrary algebraic numbers, the equation

$$N_0 e^{z_0} + N_1 e^{z_1} + \cdots + N_n e^{z_n} = 0$$

can hold only if all the N's are equal to 0.'

Lindemann's famous theorem, that π is a transcendental number, clearly follows from this result, since if π were algebraic, the famous equation of Euler $e^{\pi i} + e^0 = 0$ would contradict the given statement. Weierstrass' subsequent letters from the summer of 1882 are full of personal information. On 26 July he wrote (Kochina 1973, p. 102):

If one of my auditors, Molk, presents himself to you sooner or later, I beg you to receive him warmly. He is an Alsatian by birth, but his parents have opted for France. He studied two years in Paris and now two years here. . .From him you will be able to learn much that will be of interest to you.

Again, on 3 August he wrote (Kochina 1973, p. 102):

Your last letter contains at least something relatively happy: You have your little Sonia with you again, and I consider that very good. You will now be able to devote yourself calmly to your work or, what will perhaps do just as much good, to your recreation. . .

Perhaps Herr Molk has already presented himself or will do so in the next few days. I gave him a special copy of my Calculus of Variations for you. He can also give you a note of mine concerning Lindeman's work on π, which will perhaps interest you. . . .

It will appear in the course of the Kovalevskaya–Mittag-Leffler correspondence that Weierstrass overestimated Molk's qualities as a courier.

5.6.   The Year 1883

At our last mention of Vladimir he was returning to Russia to try to straighten out the family finances (which were already well straitened). Things did not go well for him. His position at the University of Moscow was threatened by his relation with the Ragozin company. The Ragozins had done their work very well. Vladimir was too far in debt to them to think of leaving the company and had no power to prevent further chicanery on their part. Faced with the wreck of his personal life and almost certain indictment in connection with an investigation of the company, Vladimir committed suicide on 27 April 1883 by inhaling a bottle of chloroform.

When Sonya heard of Vladimir's death, she locked herself in her room and refused to eat or allow a doctor near her. Fortunately she had the support of friends. When Kovalevskaya lapsed into a coma after five days of starvation, the doctor was able to feed her and gradually restore her health.

When she was at last ready to face life again, Kovalevskaya was confronted with enormous tasks. Most pressing was the need to finish her research into the Lamé equations so that she could take up the position which Mittag-Leffler was arranging for her in Stockholm. This position was now her only hope of financial security. In addition she could not forever leave her daughter in the care of her brother-in-law. Finally, she felt a moral obligation to untangle the financial mess left by Vladimir and to clear him of the charges against him.

In the summer she traveled to Berlin to see Weierstrass again. Evidently Weierstrass gave her either advice or encouragement. Despite her distracted state of mind, she was able to finish her research. By 27 August Weierstrass was able to write to her from his holiday resort in Grande Rive, Savoy (Kochina 1973, pp. 103–104):

> On the second day of my visit here (7th of this month) I wrote Mittag-Leffler a 5-page letter, of which four pages were about you. First I reported to him on your work, not as thoroughly as in your manuscript, but so that he can get a clear picture of what has been achieved. Then I expressed my view on the way in which you should begin your teaching in Stockholm, so that your efforts will not be hindered by any mishaps. I received a reply from him only yesterday evening and today I have already written him again. He is in complete agreement with me that you must begin with a Privatissimum in the strictest sense of the word, at most two hours per week in the first semester, so that you will have time for careful preparation of your lectures. He will provide you with students. He thinks you should be there by 15 September and even proposes that you meet him and his wife in Helsingfors around the tenth. . . .

When this letter reached her, Kovalevskaya was in Odessa at the home of her brother-in-law Alexander Kovalevsky. She had also recently attended a Congress of Natural Scientists and Physicians similar to the one in 1880 where Mittag-Leffler had been in the audience. At the Odessa conference she spoke on her solution of the Lamé equations. Following Weierstrass' advice, she immediately wrote Mittag-Leffler a letter dated 28 August (which would have been 9 September to Mittag-Leffler):

Dear Sir,
    I have finally succeeded in finishing one of the two projects which I have been working on these past two years. My first desire, as soon as I arrived at a satisfactory result, was to communicate it to you, but M. Weierstrass, with his usual kindness, took the responsibility of informing you of the results of my research, while waiting for it to be written in a form suitable for publication. I just received a letter from him telling me that he has now written to you on this subject and that you, for your part, dear Sir, responded by showing your usual good will toward me and requesting that I come to Stockholm as soon as possible to begin a course privatissimum. I cannot tell you, dear Sir, how grateful I am for the friendship which you have always shown and how happy I am that soon I shall be able to begin a career which has always been the object of my dearest desire. However I think I should not conceal from you that in more than one respect I feel very unprepared for the duties of a docent and sometimes I begin to doubt myself to the extent of fearing lest you, dear Sir, who have always been disposed to judge me with much good will, be very disappointed when you see more closely what I am capable of.
    I am so grateful to the University of Stockholm, which is the only European university willing to open its doors to me, that I feel disposed already to attach myself to Stockholm and to Sweden as to my native land; and I hope that when I come there, it will be to spend many years and to find there a second homeland. But that is precisely why I would like to come only when I feel I have merited the good opinion which you seem to have of me, and when I could hope to produce a favorable impression. I even wrote to Weierstrass today asking if he doesn't think it more prudent for me to spend another two or three months with him so as to absorb his ideas better and to fill in the gaps which may still exist in my mathematical education. I do not know, dear Sir, if Weierstrass has told you that this year he applied for a leave of absence for the winter. He has not yet received a reply from the minister; his own plans are thus not yet definite. But in case M. Weierstrass returns to Berlin at the end of October, I think it would perhaps be good for me to arrange to spend the two months of November and December in Berlin and not to come to Stockholm until 1 January 1884. The two months in Berlin would be extremely profitable for me in every respect. For on the one hand I could question Weierstrass about many points of his theories, which are not yet so clear that I would venture to expound [vortragen] them. On the other hand I would have many contacts with young mathematicians who are finishing their studies there or beginning the career of docent, with many of whom I am quite well acquainted after my recent stay in Berlin. I could even arrange a mutual exchange of mathematical communication with two or three of them. For example I would undertake to explain the theory of transformations of Abelian functions to them, which they do not know and which I have studied more thoroughly. This would give me a chance to practice in exposition [Vortragen], which I have not done at all previously; and I would arrive in Stockholm more sure of myself. But all this depends on M. Weierstrass. If he does not return to Berlin, it is naturally useless for me to go there; and in that case I shall come to Stockholm as soon as I can leave Russia and try to do the best I can. I am counting on you, dear Sir, on

your good advice and support to give me the courage to undertake a task so new for me.

In any case I shall be obliged to spend two or three weeks more in Russia to arrange my personal affairs. Unfortunately it will be absolutely impossible for me to work during that time, which I shall try to make as short as possible. I have thought of choosing as the subject of my course in Stockholm the theory of linear differential equations. I would begin with the proof (as given by Weierstrass) that every system of differential equations possesses a system of regular integrals in a neighborhood of any non-critical point. Then I would cover the investigations of Fuchs, Tannery, and Poincaré on linear differential equations. I know the literature on this subject quite well. Do you think my choice is a good one to begin with? Or would you advise me to choose something else? . . .

Last week there was a Congress of Natural Scientists and Mathematicians here in Odessa like the one in Petersburg where I had the pleasure of seeing you, dear Sir. Unfortunately this time the mathematics section was very badly represented. Neither Chebyshev nor the other professors of Petersburg and Moscow came. The only mathematician of note to attend was M. Ermakoff of Kiev,[5] whose work on convergence of series you may know. I also took advantage of the opportunity to communicate the results of my recent research. I had to speak for a rather long time and very much *in extenso,* as I had to explain the entire method which I used. Besides, my audience consisted more of physicists than mathematicians; for them the most important part of my research was that it allowed the rejection of the hypothesis of a weightless ether for the case in question, and provided the possibility of considering only oscillations of a material medium with a completely arbitrary density and distribution of matter. . . .

The letter just quoted conveys a good sense of the hopeful yet diffident mood of Kovalevskaya at this time. Her financial worries were far from over; indeed for the rest of her life she would be paying various claimants who considered that she owed them money. Her enthusiasm for life nevertheless shows through. It was the only thing strong enough to overcome her doubts and carry her into a new country, where she would have to learn a new language and find a new position in society for which there were no precedents.

The letter just quoted also shows that Kovalevskaya was already thinking of ways in which she might contribute to the intellectual life of Stockholm. She realized the importance of communication in mathematics and would attempt to channel ideas from her acquaintances in Paris, Berlin, and Moscow to her colleagues in Stockholm, and vice versa. This facet of her career became one of her most important roles and must have endeared her to Mittag-Leffler, who was keenly interested in making mathematics an international enterprise.

The uncertainties mentioned in the letter were resolved rather quickly. Weierstrass did not return to Berlin. In a month or two Kovalevskaya managed to establish Vladimir's innocence and straighten out his estate. Then,

---

[5]Vasily Petrovich Ermakov (1845–1922). He was especially interested in the pedagogy of mathematics.

leaving her daughter with Julia Lermontova, on 17 November 1883 she sailed to Stockholm on the steamship "Express."

5.7.   Stockholm

Mittag-Leffler met Kovalevskaya at the boat and took her to his apartment to introduce her to his wife Signe. For the next six weeks she lived with the Mittag-Lefflers while she got settled in Stockholm and prepared herself to teach. Even with the friendliest of new acquaintances this time must have been difficult for her. She was forced to converse in German, which was her third language, since most educated Swedes knew German better than French. In addition she must have been the first to encounter a problem which is still a source of awkwardness for many women professors today. Having her profession in common with her colleagues and her gender in common with their wives, she was forever in a kind of limbo, not belonging completely at either end of the room on social occasions. The situation was exacerbated, no doubt, by the fact that at first she was not established as a full-fledged faculty member, being completely untried in the lecture hall.

On 30 January 1884 Kovalevskaya gave her first lecture, in German. (She had offered her students French and German, and they chose the latter.) Her plan, mentioned in the letter just quoted, to lecture on linear differential equations had been modified when she learned that Mittag-Leffler had already announced such a course. Instead she gave a course on partial differential equations. She quickly overcame her shyness and proved herself an excellent lecturer, which was fortunate, as she did not receive a salary from the university but had to collect it from her students. (Among those students was a young man named Ivar Bendixson, whose work is nowadays as well remembered by mathematicians as Kovalevskaya's.) At the end of the spring term she was enthusiastically toasted by her students and presented with a framed photograph of herself. Having finished her probationary term successfully, Kovalevskaya left for a holiday in Russia and Berlin, but not without advice and work assignments from Mittag-Leffler. First of all, he advised her to bring Fufa to live with her, so as to make herself fully respectable in Stockholm society, which was rigidly conventional at the time. Second, he wanted her to get some papers for the *Acta Mathematica*, her own, of course (which had not yet been published) but any others she could locate.

Evidently the advice to bring Fufa to Stockholm conflicted with the principles of child rearing alluded to in the letter from her brother-in-law quoted above. On 17 May 1884 she wrote to Mittag-Leffler:

> I am prepared to submit to the women's tribunal of Stockholm on anything involving small matters of life. But in serious matters, especially when not only my own well-being, but also that of the child is involved, I think it would be an unforgivable weakness on my part if I allowed myself to be influenced even slightly by the wish to *appear* a good mother in the eyes of the old biddies of Stockholm.

Fufa remained in Russia for two more years.

The assignments Kovalevskaya took on while working for the *Acta Mathematica* brought her into contact with mathematicians from all over the world and caused her letters to be filled with fascinating mathematical gossip. I have been strongly tempted to translate and publish the entire correspondence, though little of it is significant for the history of mathematics. Perhaps the following samples will suffice. Regarding her still-unpublished memoir on the Lamé equations, she wrote to Mittag-Leffler on 1 June 1884:

> Yesterday I began working on my memoir. Weierstrass is not eager to publish his, and proposes that I communicate the contents under his name in my memoir. I admit that this has its good side; however, nothing is yet decided, and I am afraid that after I have written it all, I shall have to start over from the beginning. . . .

Kovalevskaya was obviously afraid Weierstrass would change his mind at the last moment and publish the memoir himself, forcing her to rewrite her own part completely. As it happened, he did not change his mind. On 7 June she wrote:

> Weierstrass has just reread the work "on Abelian integrals of third rank which reduce to elliptic integrals by a transformation of second degree," which I wrote in 1874, and he finds it completely suitable for publication without any correction. I am taking the liberty of sending it to you. Do you think you could put it in the *Acta* while waiting for my new memoir, which I hope to be able to send you within a month? It would perhaps be a good idea to publish everything that I can just now. I have also located my manuscript on Saturn's ring and I am going to send it to Gyldén.
>
> Excuse me for sending you this manuscript written in Gothic characters, but I don't think you will have any great difficulty reading my German hand writing. It is the same with my [spoken] German: It is more comprehensible for the foreigner precisely because it is so little like German. . .
>
> I have had very bad luck since I arrived in Berlin. I caught a chill en route and then contracted a cold accompanied by fever, which has prevented me from visiting Kronecker and Kirchhoff. Today is the first time I have even been able to receive any of my young friends. . . . Today fortunately I hope to be able to dine at Mme. Borchardt's accompanied by M. Fuchs and his wife.
>
> I thank you very much for sending me your memoir. Weierstrass also is about to read it just now. He will be able to use your proof of Laurent's theorem in his course this summer. I loaned my copy of your memoir to M. Runge, and I promised to send you his memoir on the development of an arbitrary function in rational fractions which he finished last fall. Since his point of departure (the Cauchy integral) is completely different from yours, I think it would perhaps be interesting to publish it; for the subject is truly of the greatest importance and two proofs would not be too many. . . .

The following undated letter obviously follows very closely the one above. Explanatory notes follow.

I shall begin by answering the questions you posed in your recent letters.

1. Weierstrass has read your memoir on Laurent's theorem; as for the large memoir, he began reading it only yesterday. He told me yesterday evening that he is not finding it easy reading, but I suppose that is only for the following reasons: Weierstrass does not read French as easily as German and besides he is at present in a state of chronic lassitude and exhaustion. But consider also that at his age [nearly 69] he gives 6 courses per week and in addition a 2-hour course at the seminary every two weeks. In the latter course Weierstrass is giving an exposition of certain theorems of uniform functions. He told me that he intends to give an exposition of your large memoir as well when he arrives at those questions. So you will certainly have his opinion very soon. I tried to talk with Kr[onecker] about your memoir but without success. Kr[onecker] showed himself more than reserved on the subject; he evidently did not want to say anything. He remarked, with a certain acerbity, only that you have evidently taken a great deal of trouble to attach the investigations of Cantor to a subject with which they have nothing in common. As for Fuchs, he is now absolutely crushed beneath the weight of his new dignity, and he reads nothing but what is absolutely necessary for his courses. I have seen him 3 times at dinners, but found it absolutely impossible to get him to talk mathematics. Every time I began to tell him something, he said nothing but an occasional "hm! hm!" and when I asked him questions, he sighed, adopted the air of a martyr and appeared to be saying to me, "In the name of heaven will you let me digest my food in peace."

We have talked a lot with Runge about your memoir and his, which he had already handed to Weierstrass last autumn. As I have already told you, you and he arrived at almost identical results, but by completely different routes. Runge will send you his memoir tomorrow or Saturday, and I think it will be very interesting to publish it if only to show up the difference and at the same time the analogy between research based only on series developments [Mittag-Leffler's] and research based on the Cauchy integral [Runge's].

Speaking of Cantor's research, one of the most beautiful applications I have seen of it was made by Hurwitz, to prove that a function of *several* variables whose only singularities are poles is necessarily a rational function.[6] Weierstrass had often announced this theorem, but he had never proved it anywhere. Do you have any acquaintance with Hurwitz's work? From what I hear in Berlin he is a very young geometer (25 years old) of very great talent. He has just been named professor extraordinarius in Königsberg. He has already produced some very pretty things. His inaugural dissertation on the modular function is very beautiful, but Weierstrass attaches even more value to his work on 2n-periodic functions which have a fundamental system of n real periods. That is one which we positively must get for the *Acta*. But I don't know how to do it, seeing that he isn't in Berlin.

I believe I have already written you that Minkowski is also not in Berlin at the moment. He didn't produce anything this winter, but actually that is a very good thing, for he has much still to learn. Just imagine, he is only 20 years old!

It is extremely curious that no one here has heard anything about Söderblomme. His name is not listed among the students of Weierstrass or Kronecker. Probably he is not in Berlin.

---

[6]Compare the footnote in Weierstrass' *Werke* III; 113, where Weierstrass says that a function which has only algebraic singularities must *be* algebraic.

2. I questioned Weierstrass on Krazer's memoir.[7] He told me that he
hasn't read the whole thing, but that he is certain that it contains nothing very
interesting, and that besides he has very little hope for Krazer, whom he knows
personally. But Krazer's master Prym is also not a very profound mathematician. So I suppose that you will not lose much by relinquishing the new
memoir by the same author.

Today I received the new memoir of Wildtheiss, and I have already read
half of it. The formulas are certainly correct but do not seem interesting to me.
I shall try again to have Weierstrass' opinion on the subject.

3. Next semester Kronecker will give a large course on algebra.

4. In regard to the question of the prize [the Oscar II prize] Weierstrass
has promised me that he will write you his opinion on that in more detail as soon
as he receives a letter from you. I did not inform him of what you wrote me
in the letter before last with regard to the choice of jury, for I was sure in
advance of his complete disapproval. Indeed I believe that in this way the thing
presents many practical difficulties. Just consider how one could hope that four
famous mathematicians, Weierstrass, Hermite, Cayley, and Chebyshev would
ever agree on the merits of a memoir. I believe it is certain that each of the four
would refuse to become part of the jury as soon as he learned the names of the
other three. As for Weierstrass, I am so sure of this that I didn't even venture
to talk to him about it. In general Weierstrass thinks that it will be quite difficult
for the jury to agree when they have no opportunity to talk face to face. To do
it by mail is considerably more difficult; and at bottom, why *would* these old
gentlemen take so much trouble for us? There, I fear, is a very great difficulty!
As for the honor, quite the contrary, each of the four that you named will be
outraged that you chose the others along with him.

5. But at last I can inform you of something which I hope will give you
pleasure. Weierstrass has promised me in all seriousness to come to Stockholm
next fall and give a course of six seminars. As for the reimbursement, he
insisted that he would not accept the 5,000 kronor but that he would perhaps
have his journey paid for and his living expenses in Stockholm. I believe this
is a completely serious promise and that Weierstrass wants very much to come
to Stockholm when I become a professor there. [Kovalevskaya had not yet been
reappointed for a five-year term as professor and was still docent.] He might
have done so this year if it had not been for the illness of his sister Clara, which
forced him to go to Switzerland. At bottom Weierstrass is much more anxious
than I for my appointment. Since I gave up all hope of getting into courses here
this semester, I await it much more philosophically. Weierstrass is very anxious
and wants very much to present me here as a professor. He talks about it every
time he sees me, which is at dinner nearly every day.

I have already written you that the *Illustrierte Zeitung* has published my
biography and my portrait (an awful one) but there was no question of professor, only of privat-docent. Can you believe that a certain Monsieur Fried-
länder took it on himself to address a poem on the subject to me. (It was in one
of the two letters which you forwarded to me from Stockholm) The other was
from the *Wiener Residenz Blatt*, which asks my permission to do the same, i.e.
publish my portrait. I haven't given any answer yet. I am sending you the letter.
Do you think it would be appropriate to reply?. . . .

This letter, brimful of enthusiasm, shows how much Kovalevskaya's psychological state had improved in just a few months. In her next letter (1 July

[7] Adolph Krazer. Despite Weierstrass' slur he had a respectable career.

1884) to Mittag-Leffler she expressed her gratitude and joy that her appointment as professor had been approved. In order to secure it Mittag-Leffler had been forced to do some political trading. He agreed to withdraw his objections to the appointment of two young men in return for Kovalevskaya's five-year appointment (Koblitz 1983, p. 187). She concludes the letter with some personal information:

> As for the biographical dates about me which you requested, I think you know them all. I was born in 1853, married in 1869 and the same year accepted as a student of mathematics in Heidelberg, where I also took physics courses from Kirchhoff and Helmholtz. In 1871 I came to Berlin where I studied privatissimum with Weierstrass until 1874, when I obtained the doctor's degree *in absentia* at Göttingen. That is all of any interest about me that can be communicated. . . .

The reader will note that Kovalevskaya was fibbing about her age here, making herself three years younger than she actually was, and moving her marriage back one year, so that she would seem to be 16 when she was married.

Thus after only one term of teaching Kovalevskaya was established as a mathematician and as a resident of Stockholm. In her early enthusiasm for her new position she had made inquiries about becoming a citizen of Sweden, but eventually decided not to bother. Though Stockholm contained some very good scholars, like most cities it suffered when compared with Paris and Berlin, where Kovalevskaya was well known in mathematical circles and felt quite at home. To stimulate mathematical activity in Stockholm, Mittag-Leffler had obtained the cooperation of the best Scandinavian mathematicians on the editorial board of the *Acta,* such as Bjerknes, Sylow, Lindelöf, and Lie. He also conceived the idea of an endowed chair for distinguished visiting professors from abroad. Against the advice of Weierstrass, who pointed out that the Prussian ministry of education would not allow a professor to take a leave of absence in order to give courses in another university, Mittag-Leffler and Kovalevskaya began seeking just such a mathematician. Mittag-Leffler suggested Runge, who was mentioned favorably in the letters quoted above. Somewhat surprisingly Kovalevskaya demurred, saying she now found him very vain and disagreeable. (Her pique may have been caused by Runge's gratuitously pointing out to Mittag-Leffler that Cauchy had anticipated some of Kovalevskaya's work on differential equations.) Nevertheless Runge did come to Stockholm that autumn (though not as a distinguished visiting professor), and Kovalevskaya managed to cooperate with him. Mittag-Leffler took Kovalevskaya's advice and printed Runge's paper in the *Acta Mathematica.* This paper (1885) contains the famous theorem of complex analysis nowadays known as Runge's theorem. When Runge returned to Berlin he sent Kovalevskaya (17 November) a list of misprints in her article on Lamé's equations, evidently too late to be corrected, since the recommended changes were not made.

A seemingly trivial event at this time turned out to be the harbinger of very important events in Kovalevskaya's future. After returning from a brief stay in Södertälje, near Stockholm, she went to the Post Office and found a package addressed to Monsieur le Professeur Kowalevski. Having, as she said, no claim to such a title, she refused it and explained at length to the postmaster that letters addressed to Madame Professor Kowalevski should be sent to her in Södertälje, while those addressed to Monsieur Professor Kowalevski should be kept. Nevertheless, upon her return to Södertälje she found the same accursed package waiting for her. It is not known whether the addressee, Professor Maxim Maximovich Kovalevsky ever received the letters, but on this occasion he does not seem to have met Kovalevskaya. He had, however, met her briefly two years earlier among her radical friends in Paris (Kochina 1981, p. 149).

M. M. Kovalevsky (1851–1916, and at most distantly related to Vladimir) was a very learned writer on social history and law, the author of many highly regarded books. Like many Russian jurists, he was a socialist and a friend of Karl Marx, who made extensive notes on Kovalevsky's book on the development of the economic system of India. Both Marx and Engels corresponded with Kovalevsky. He was to play a very important role in Kovalevskaya's life from 1887 on.

5.8. The Academic Year 1884–1885

In October Kovalevskaya, apparently on a whim, asked the Russian ambassador in Stockholm, Shishkin, to arrange for her to be received by King Oscar II. Shishkin made the arrangements through a certain Baron Hochschild, and Kovalevskaya was officially received on Tuesday 28 October. After the interview she wrote to Mittag-Leffler, "I have just visited the King, who was very nice. I had to give him a whole course on the theory of Helmholtz's *Obertöne*; he seemed very interested!!!!. . . ."

All summer long Kovalevskaya had been trying to get her memoir on Lamé's equations ready for publication and, as mentioned, all summer long Weierstrass had been unable because of exhaustion to do his part. Finally in response to pleas from her which must have become more and more strident, he wrote her a letter of explanation of 13 September, in which he suggested that she send her part of the memoir for publication first, to be followed by his part (whenever he should get around to doing it, one supposes). This letter has been frequently misconstrued by writers on Kovalevskaya, who did not understand the context of the apologies Weierstrass made in the letter (cf. Chapter 9). Weierstrass was evidently too exhausted at the time to look at Kovalevskaya's half of the paper, though he did send her on 17 October a list of suggestions regarding his own part, which were carefully followed when the paper was published.

During 1884–1885 Kovalevskaya took an active part in the preparations to honor Weierstrass on his 70th birthday, which was to be 31 October 1885.

During the Christmas vacation she traveled again to Berlin, where she received, as she later wrote to Mittag-Leffler, "a Christmas present from your sister, an article by Strindberg in which he proves as clearly as $2 \times 2 = 4$ how pernicious, useless, and disagreeable is such a monstrosity as a woman professor of mathematics. . . ." One would, of course, expect such a view from the author of *Miss Julie*. A more practical Christmas present came from Hermann Amandus Schwarz in the form of a request for help on a problem of calculus of variations (Kochina 1980). He needed to know more about a certain partial differential equation. Kovalevskaya's reply, if any, has not been located.

In the spring of 1885 Kovalevskaya was on the periphery of an acrid controversy occasioned by Mittag-Leffler's following her advice in regard to the composition of the jury for the Oscar II Prize. Kronecker, who was insufferably vain to begin with, was grievously offended that he had been passed over when the committee was made up and outraged that the committee of Weierstrass, Mittag-Leffler, and Hermite had dared to pose as one of the questions the study of the algebraic relations connecting two Fuchsian (automorphic) functions having the same automorphism group. Believing himself to be, not the world's greatest expert on algebra, but the world's *only* expert on the subject, he wrote an angry letter to Mittag-Leffler, which he began by refusing to fulfill his earlier promise to recommend a gynecologist for Mittag-Leffler's wife and ended by threatening to inform King Oscar II that he had long ago proved the impossibility of obtaining such results.

Kovalevskaya, who had been behind the scenes during the entire controversy, was amused by Kronecker's unbridled egotism. Mittag-Leffler tried to smooth things over by explaining that he had wanted to honor Weierstrass because of his advanced age. When this explanation reached Weierstrass, he in turn was offended; and the whole project continued in an atmosphere of bad feeling. Kovalevskaya was able to remain loyal to her true friend (Weierstrass) without offending his adversary. She had plenty of practice in such diplomacy, since she was sometimes caught between Mittag-Leffler and Hermann Amandus Schwarz, who were at daggers drawn most of the time.

In the summer of 1885 Holmgren, a professor of mechanics, fell ill; and it was necessary to find a replacement for him. After a considerable search, the faculty decided to allow Kovalevskaya to replace him temporarily. During that same summer a vacancy appeared in the Swedish Academy of Sciences, and Mittag-Leffler immediately tried to amend the charter of the Academy by replacing the world "män" with the word "personer" (cf. Webster 1894), so that he could nominate Kovalevskaya. Realizing that her appointment would be at best a pyrrhic victory for her supporters because of the ill will created, Kovalevskaya persuaded Mittag-Leffler to abandon the project.

The Weierstrass jubilee went off with éclat in the fall of 1885, though Kovalevskaya did not attend, being busy with her courses. Weierstrass was pleased with the ceremony and sent Kovalevskaya a very cheerful report on it. On this 70th birthday he was inspired to write a poem, which he also sent her.

5.9.   The Academic Year 1885–1886

The next few months were among the most productive of all Kovalevskaya's mathematical career. She finally clarified the situation in regard to the use of theta functions to solve the Euler equations for the rotation of a rigid body about a fixed point: They can be used only in the already-known cases of Euler and Lagrange, and in one new case (now known as the Kovalevskaya case) which she had discovered. Her undated communication of this result to Mittag-Leffler (evidently early in 1886) read as follows:

> Dear Sir,
>   I thank you for your invitation for tomorrow, and I shall come with pleasure.
>   It is a question of integrating the following differential equations.

$$A\frac{dp}{dt} = (B - C)qr + z_0\gamma' - y_0\gamma'' \qquad \frac{d\gamma}{dt} = q\gamma'' - r\gamma'$$

$$B\frac{dq}{dt} = (C - A)rp + x_0\gamma'' - z_0\gamma \qquad \frac{d\gamma'}{dt} = r\gamma - p\gamma''$$

$$C\frac{dr}{dt} = (A - B)pq + y_0\gamma - x_0\gamma \qquad \frac{d\gamma''}{dt} = p\gamma' - q\gamma.$$

> Up to now they have been integrated only in 2 cases: (1) $x_0 = y_0 = z_0 = 0$ (the case of Poisson and Jacobi); (2) $A = B$, $x_0 = 0$ (the Lagrange case).
>   I have found the integral also in the case $A = B = 2C$, $z_0 = 0$, and I can show that these 3 cases are the only ones in which the general integral [i.e. the solution for every set of initial values of the variables] is a single-valued analytic function of time having no singularities but poles for finite values of t. . . .

Thus after four years of prospecting Kovalevskaya had finally struck gold. Her result, as announced here, consisted of two parts: (1) a new special case where the equations could be completely integrated; (2) a proof that no other cases remained in which the solutions were meromorphic functions of time. Both parts are mathematically significant. Kovalevskaya's new case represented physically a *nonsymmetric* body, and the rotation of such a body under the influence of gravity is extremely complicated. The previously studied cases cited by Kovalevskaya were for a body unaffected by external forces (what she called the Poisson–Jacobi case) and for a body with two of its principal moments of inertia equal and its fixed point on the third principal axis. This case applied mostly to symmetric bodies. It is significant that she attributed the first case to Jacobi. Actually, as she knew very well, it was Euler who first studied this case. Jacobi's contribution had been to use his theta functions to find closed-form integrals of the differential equations.

The negative half of the theorem was also important, since it would prevent further attempts to solve these equations using the techniques of

Abelian functions. We see then that by the spring of 1886 Kovalevskaya had solved "in principle" as much of the problem as anyone could reasonably hope to do. The rest was a matter of working out details, but a considerable amount remained to be done.

In fact what remained was to get these results written in a form suitable for communication. As a true student of Weierstrass, Kovalevskaya felt obliged to do more than simply say that the integrals could be inverted using theta functions (which was a consequence of Weierstrass' work on Abelian functions). She planned to carry out the inversion in detail, and show explicitly how to express the parameters which describe the motion as functions of time. In the meantime, she knew she must let others know what she had done as soon as possible. Accordingly she left for Paris as soon as her classes were finished. What happened there can be surmised from the following two letters. The first is from Hermite to Kovalevskaya, 16 June 1886:

> Your beautiful discovery on rotation was received with the most lively interest by M. Halphen, M. Darboux, and M. Picard, and I am instructed by M. Halphen to beg you to inform him when you would do him the honor of receiving him so that he can discuss it with you. . . .

The second is from Kovalevskaya to Mittag-Leffler, 26 June 1886:

> M. Bertrand evinces a very extraordinary benevolence toward me. Just imagine what he has thought up. Next Monday these gentlemen are to meet and propose the topic for the Grand Prize of the Academy for the year 1888. Bertrand had the idea of proposing as a topic precisely the problem of the rotation of a rigid body. That way I shall have a chance of obtaining the prize. You can imagine how this idea tempts me. Yesterday Hermite, Bertrand, Camille Jordan, and Darboux, who are all members of the committee, discussed the project with me. They had me tell them in detail the results of my work, and they seemed to find it so interesting that they believed it has a good chance of winning. The only disadvantage is that in that case I shall have to postpone publication until 1888. I shall have to submit it to the Academy, and the decision will be made in 1888. You can imagine how this project attracts me. Only in that case it is impossible for me to communicate my work at Christiania this year. That would be much too risky. . . .

As things turned out Bertrand et al. arranged a competition suitable for Kovalevskaya's memoir, only not for the Grand Prize, as she had anticipated. The competition was proposed for the less prestigious but equally lucrative Bordin Prize. Bordin Prizes were awarded in many areas, such as Botany, Chemistry, Physics, and Mathematics. The competition was announced in Vol. CIII, No. 26 of the *Comptes Rendus* (1886) on p. 1395.

PRIX BORDIN

(Question proposed for the year 1888)

> "Improve, in some important point, the theory of the movement of a rigid body."

> The prize will be a gold medallion worth three thousand francs.
> Manuscript memoirs submitted to the competition shall be received by the Secretariat of the Institute before 1 June 1888; they shall be accompanied by a sealed envelope containing the name and address of the author. This envelope will be opened only if the memoir to which it corresponds is awarded the prize.

By the usual rules of the Academy the author wrote an identifying phrase on the outside of the sealed envelope and the same phrase on the memoir, so that there could be no doubt which envelope belonged to the memoir. Those who did not win were granted anonymity and the freedom to submit their work elsewhere.

5.10.   Life as an Untenured Professor

This pleasant interlude in Paris was followed by a somewhat less pleasant journey to Russia. Kovalevskaya's sister Aniuta had been chronically ill for a long time. In the summer of 1886 she grew much worse, and Kovalevskaya went to Petersburg to see her. When Aniuta's husband Victor Jaclard arrived, Kovalevskaya left Aniuta in his care and went to Moscow to see Julia Lermontova and–at last–to pick up her daughter. The academic year 1886–1887 therefore promised to be the busiest one yet. Kovalevskaya was going to be preoccupied by setting up housekeeping with Fufa, worrying about Aniuta, and writing up her memoir, not to mention teaching classes.

Aniuta grew worse in the autumn, and Kovalevskaya, putting her sister ahead of her career, asked for a leave of absence to go to Russia and take care of her. This request was not only refused; it was considered an impertinence. Male professors never asked for leave to care for sick relatives! It seems ironic that the conflict between career and family, which has been unfairly imposed on women, should appear even in the case of a widow.

An undated note from Kovalevskaya to Mittag-Leffler, which may have been written about this time, illustrates another problem which unfortunately still afflicts many career women today, the need to be constantly proving their devotion to the profession. As a regular faculty member (officially Professor Extraordinarius) Kovalevskaya was the first woman in modern times to assume this role. Naturally some of the smaller-minded faculty members kept a sharp eye out for any sign of unprofessional behavior on her part. The result is the following note:

> . . .Could you please tell me if you know anything about today's faculty meeting and whether you intend to go? I think that I myself must go, so as not to give Peterson and Lecke the chance to say, "Professor Kowalevski naturally didn't come because he was occupied with the bazaar. . . ."

At the Institut Mittag-Leffler I found a note to Kovalevskaya from the Russian Embassy which mentioned an upcoming bazaar. Evidently Professors Lecke and Peterson thought it humorous that a professor would get involved in such undignified activity.

At the end of the fall term Kovalevskaya returned to Russia to see Aniuta and with instructions from Mittag-Leffler to procure, if possible, a subsidy for the *Acta Mathematica* from the Russian government. On 19 December she wrote that she had spoken with a government official named Osten-Saken, who told her that the Russian economy was in no condition to make any such commitment. He counseled her to see Count Tolstoy (no apparent relation to the famous writer). Kovalevskaya evidently knew Tolstoy and told Mittag-Leffler that he would be very flattered to receive a letter from King Oscar II. Evidently the flattery did not work if it was tried, for no support ever came from the Russian government for the *Acta*.

5.11. New Distractions

In the spring term of 1887, when she should have been working full-time on her memoir, Kovalevskaya found two new interests. First her old acquaintance Maxim Kovalevsky came to Stockholm to deliver a course of lectures, having just been fired from Moscow University for disapproving of the government's policies too forcefully in his lectures. Second, Kovalevskaya conceived the idea for a play–two plays, really, one depicting the depressing way the lives of a group of young people actually turned out, and one showing the inspiring way it might have been. She inveigled Mittag-Leffler's sister Anne-Charlotte into doing the writing, since she was not a master of Swedish prose. All this peripheral activity annoyed her patron Mittag-Leffler, but he could only look on in exasperation as his protégée wasted (so he thought) her time and talent.

In the summer of 1887 a new crisis arose in Russia. As Koblitz describes it (1983, p. 202) Aniuta's husband Victor Jaclard had been a correspondent for several French newspapers and had written approvingly of the attempt by Lenin's older brother to assassinate Tsar Alexander III. Despite Aniuta's grave illness, she and Victor were ordered out of Russia. Kovalevskaya took care of Aniuta until she was well enough to join Victor in Paris, but apparently did not accompany them to Paris. A letter from Aniuta to Kovalevskaya, dated 12 August 1887, which I found at the Institut Mittag-Leffler, expresses some relief that Sonya had not come to Paris, since, as Aniuta put it, she would simply have whined all the time and made her miserable.

The fall of 1887 and the early part of 1888 were an emotionally turbulent time for Kovalevskaya. Aniuta died of postoperative complications in October. Shortly after this loss, which she felt very keenly, Kovalevskaya became interested in the Arctic explorer Frithiof Nansen and subsequently, much more intensely, in Maxim Kovalevsky. Eventually Mittag-Leffler persuaded Maxim to travel to Uppsala so that Sonya could finish her memoir.

5.12. The Bordin Competition

With Maxim gone, Sonya concentrated very hard on her work, but was
stymied by the problems she encountered in transforming the integrals. She
wrote to Weierstrass for help, but, as so often happened, he was too ill to
work. Fortunately, by the time he answered her request, she had already
overcome the difficulties on her own, and his help amounted only to a few
minor rearrangements.

The memoir was not actually finished by the 1 June deadline. However,
Kovalevskaya sent a half-ready version of her work and asked permission to
send in a revised version later. The ensuing events were vividly described in
two letters from the summer of 1888. The first is from Charles Hermite to
Kovalevskaya, 11 June 1888:

> . . . My colleagues M. Bertrand and M. Darboux, with whom I took the
> occasion to discuss the subject of the Bordin Prize, confirmed totally everything
> M. Pingerd told me and instructed me to let you know that you may, if you
> believe it necessary, send a new version of the memoir you submitted to the
> competition, since the committee of the Academy charged with examining the
> entries has complete power to decide on the deadline. And since these Academ-
> icians take vacations, of which they have great need, you may rest assured that
> they will not set it before the month of October, so that your new version or
> supplement to the first will be on time if it arrives at the Secretariat during the
> month of September or even at the end of the month.
>
> Two memoirs were submitted to the Bordin Prize competition at the same
> time as yours, one from Paris, the other from Brest. They could not be commu-
> nicated to me, since I am not part of the committee, but I do not suppose that
> they have for authors any geometers known to us. . . .

The next quotation is from an undated letter from Kovalevskaya to Mittag-
Leffler. Internal evidence puts the letter sometime during June 1888:

> . . .Last Tuesday Bertrand gave a large dinner in my honor, attended by
> Hermite, Picard, Halphen, and Darboux. Three toasts were proposed in my
> honor, and Hermite and Darboux said openly that they have no doubt that I shall
> have the prize. . . .

Since Darboux was a member of the jury, one would say that Kovalevskaya's
chances were pretty good, though there is some doubt as to the anonymity of
the competition.

Kovalevskaya submitted a revised memoir to the Paris Academy in the
late summer of 1888. The revised memoir was still not complete, but she was
satisfied with it and confident of its quality, as she wrote to Mittag-Leffler
from Wernigerode in the Harz Mountains, where she had gone to consult with
Weierstrass, Hurwitz, and other Berlin mathematicians. Her confidence was
well justified. On 24 December 1888 Mittag-Leffler received a telegram from
Paris saying, "Bordin décerné à Madame Kovalevsky. . . ." Kovalevskaya
herself was already in Paris, where she had gone with Maxim Kovalevsky to

await the outcome of the competition. The telegram (evidently from Kovalev-skaya herself) went on to say that the prize had been doubled due to the originality of the results.

The report of the event in the *Comptes Rendus* lists the jury for the competition and the reason for the committee's decision. The jury consisted of five men. The first four are listed as Maurice Lèvy, Phillips, Resal, and Sarrau. Following a semicolon the fifth member, Darboux, is listed as spokes-man for the committee. (It is not clear whether Darboux had a vote. If he did, the contest was certainly not anonymous, since he had arranged it himself after hearing Kovalevskaya describe her work in 1886. The point is not worth emphasizing, however, since it is practically impossible to arrange an anon-ymous scholarly competition under any circumstances. There is no doubt that Kovalevskaya's work on this problem was the best of its kind for many years.) What impressed the committee, and indeed the entire mathematical world, was the fact that Kovalevskaya had applied the recently developed and highly abstract theory of Abelian functions to solve a problem which arose in physics, thereby justifying to some extent the enormous amount of work expended on this theory. As the committee's report said:

> . . .The author has not contented himself with adding a result of very high interest to those which were bequeathed to us by Euler and Lagrange; he has made a profound study of the result due to him, in which all the resources of the modern theory of theta functions of two independent variables allow the complete solution to be given in the most precise and elegant form. One has thereby a new and memorable example of a problem of mechanics in which these transcendental functions figure, whose applications previously had been limited to pure analysis and geometry. . . .

5.13. Reactions

Kovalevskaya's escort during all the ceremonies, Maxim Kovalevsky, was somewhat miffed at being "Mrs. Kovalevsky," to use a phrase of Koblitz'. Relations between him and Sonya were constantly being strained after that time, probably because of his swollen and bruised ego. To say that Ko-valevskaya experienced a letdown after this triumph would be an abuse of language. She put it succinctly in a letter to Mittag-Leffler of 12 January 1889:

> . . .I am receiving letters of congratulation from all over and by a strange derision of fate I have never in my life felt so unhappy as I feel at this moment. Unhappy as a dog; no, for the dogs' sake I hope that they cannot be as unhappy as men, and especially women, can be. . . .

Kovalevskaya was in a state of total exhaustion and nervous collapse. She asked Mittag-Leffler to get her a leave of absence for reasons of health. This he did, despite the fact that her five-year contract was due to expire at the end of the term, and he was trying to get it renewed. Besides the natural exhaustion after her prodigious exertion, Kovalevskaya had a second reason for wishing to stay in Paris. Stockholm had for a long time seemed rather provincial to her. Paris offered a more active social and intellectual life and less need for conformity. In addition she was now quite enamored of Maxim Kovalevsky, who evidently could not or would not come to live in Stockholm. Kovalevskaya decided to use her new-found fame and her prize-winning paper to get a position in Paris. To obtain such a position she would need six months residency in France in order to become a naturalized citizen (of one class); and she would need a doctoral degree from a French university. She hoped the leave of absence would get her the residency and she planned to use her new memoir to get the degree.

When Weierstrass learned of her plan, he was appalled. He wrote to her in the strongest possible language, telling her that any hope of a position near the center of French academia was out of the question for a foreigner. Moreover, he said, one could not render any greater insult to the faculty at Göttingen, which had awarded her the doctoral degree than to accept a second degree in the same field from another university (Kochina 1973, p. 148).

Although she eventually gave up her plan to leave Stockholm, Kovalevskaya was now an international celebrity. In May 1889, with recommendations from Bjerknes, Hermite, and Beltrami, the University of Stockholm made her (full) professor for life. A recently published portion of the Hermite-Stieltjes correspondence (Cahiers du séminaire d'histoire des mathématiques 4, 1983) shows that Hermite was flustered by the request from the University for an evaluation of Kovalevskaya's work. He asked Stieltjes to write the evaluation for him on everything except the memoir on the rotation problem. Stieltjes in turn limited his comments to the papers on partial differential equations and Saturn's rings. He suggested Picard as a suitable reference for the paper on reduction of Abelian integrals.

5.14.   Final Days

For the first nine months of 1889 Kovalevskaya used her leave of absence to travel to various resorts, occasionally seeing Maxim Kovalevsky. This was the year of the Paris Exhibition, and the Eiffel Tower was attracting many visitors. Kovalevskaya entertained the many people among her acquaintances who came to Paris. Among these were Mittag-Leffler and his wife, who brought Fufa along for one of her rare visits with her mother. Also among them was Kovalevskaya's distant cousin Lieutenant-general Kosich, who was most impressed with the cousin he had not seen for many years. Upon his return to Russia he decided that she must be brought home. He sought a position for her in Russia, but unfortunately the barriers to women were still in place. No position commensurate with her talent was open to a woman. However, upon the recommendation of Chebyshev and others she was made a corresponding member of the Academy of Sciences. Her nomination as a *full* member the following year was not successful, as we shall see.

When Kovalevskaya returned to Stockholm in the fall of 1889, she had no really new mathematics to show, though her memoir on the rotation problem had been expanded. In fact she wrote and published two variations on this memoir to clarify points which were left obscure in her haste to meet the deadline for the Bordin competition. Her autobiography of her childhood appeared in Swedish translation and has been highly regarded ever since, both for its psychological insight and for the portrait it presents of Russian village life. She was expected to give a paper at a mathematical conference which was held at the University. To meet this expectation, she apparently dipped into some unpublished parts of the dissertation she had written 15 years earlier to find an application of her theorem on partial differential equations. (Such, at any rate, is my interpretation of the facts. The evidence in support of this claim will be presented at the end of Chapter 8.) For one of her variants of the paper on the rotation problem she received a prize from King Oscar II. This memoir appeared in the *Acta Mathematica* before her original memoir was published in Paris.

The principal justification for the second memoir was to give the details of her proof that she had found the last case in which the solutions of the Euler equations are single-valued functions of time. Her argument left a small gap, however, even though her claim was correct. This paper was carefully read by the brilliant young mathematician A. A. Markov, who proclaimed the "mistake" very loudly at a crucial time for Kovalevskaya. The academician Bunyakovsky had died early in 1890, and Kovalevskaya had hopes of being elected to fill his vacancy. She journeyed to Petersburg in May 1890 and spoke with Chebyshev and others about the possibility. Chebyshev was encouraging, but the project fell through, nevertheless. As she wrote to Mittag-Leffler:

> Markov has said publicly that my work on rotation (the prize-winning memoir) is full of the grossest errors. When challenged to point out the errors, he replied rudely that he would do so, as soon as certain academicians who have proposed me for membership will trouble themselves to actually read my memoir!!! At the time none of the mathematicians present was disposed to argue. Several weeks later Imshenetsky asked him about his quarrel with me. By then Markov had managed to get himself elected extraordinarius, and he graciously allowed that the situation with regard to my memoir was not so bad as he had at first believed, but still there was little of value in it. From many sources I have heard the pleasant tale that the non-mathematical members of the Academy are said to be convinced that Markov has completely demolished my memoir. I conveyed to Chebyshev and Imshenetsky my demand that they take some *public* occasion in the Academy to clarify the matter. They answered me evasively that Markov is still a young man, and one cannot be too strict with him. I do not believe they will do anything, and don't know what I can do myself. I will not leave Petersburg without seeing the Grand Duke. . . .

Although the Grand Duke was encouraging, it was clear that Kovalevskaya had no chance of being elected to the Academy. An attempt was made to complete Kovalevskaya's argument by a young student named Appel'rot, who soon produced a paper (Appel'rot 1892) which Markov considered inadequate. Markov then assigned the problem to Lyapunov, who proved decisively that Kovalevskaya's claim was correct (Lyapunov 1894; also in Kovalevskaya's Raboty, pp. 286–304).

Even though Kovalevskaya did not gain membership in the Academy on this trip, she did give an autobiographical account of her mathematical career which was later printed (cf. Stillman 1978, pp. 313–329). After leaving Petersburg she dropped in on Weierstrass in Berlin, then spent the summer traveling with Maxim Kovalevsky, who had just spent the year in Oxford delivering the Ilchester lectures (Kovalevsky 1891). The two quarreled and reconciled repeatedly.

Although Kovalevskaya wrote in September 1890 that she was happily immersed in her work (Koblitz 1983, p. 202) and told Poincaré that she was sending a letter to Hermite describing new results on the rotation problem, no such results have been found among her papers. (M. Pierre Dugac tells me that the letters Hermite received were destroyed in a fire shortly after his death.)

After the fall term in 1890 she traveled to Genoa to spend the holidays with Maxim. Somewhat disorganized on the return journey, she arrived in Copenhagen without any Danish money to tip a porter. As a result she had to carry her own bags in a driving rain. When she arrived in Stockholm, she was ill, but managed to teach her first class of the new term on Friday 6 February, before taking to her bed. On Monday she seemed better and described her plans for more work on the Euler equations to Mittag-Leffler. (See Mittag-Leffler 1892. Unfortunately he was unable to recall what she said, beyond the fact that it was a plan to use Weierstrass' theory of Abelian functions.) Mittag-Leffler thought that this work would be better than anything else she had done. However, time ran out for Kovalevskaya with terrible suddenness. She died early Tuesday morning, 10 February 1891.

The news of Kovalevskaya's death saddened people throughout the world. Mathematicians, artists, reformers, and intellectuals sent telegrams and flowers. Mittag-Leffler's brother Fritz wrote a poem entitled "Vid Sonja Kovalevski's Graf." Maxim Kovalevsky and Mittag-Leffler gave eulogies which recounted her many achievements. Her grave in Stockholm's Nya Kyrkogarten (now the Norra Begravningsplats) is marked by a beautiful Russian cross with the inscription in prerevolutionary Russian orthography: From her Russian friends and admirers. It is frequently decorated by those who still honor her memory. Yet the most poignant tribute, certainly the one which would have meant most to her, has long since withered away. It was a wreath of white lilies sent to her funeral with the message. "To Sonya, from Weierstrass."

CHAPTER 6
The Lamé Equations

6.1. Introduction

The ultimate origin of the problems which led to Kovalevskaya's paper (1885a) is in the wave theory of light, specifically the mathematical explanation of refraction. We shall first look at a simple explanation of refraction in this introductory section, then examine the applications of this principle to more complicated cases by consideration of the work of a few of the scientists who developed the theory. These few aspects of the theory suffice to give the background for the problem studied by Kovalevskaya. For reasons which will be apparent at the end of the chapter, Kovalevskaya's work in this area has received very little attention.

When light travels across two different media with two different velocities, refraction occurs at the interface, that is, the wave front changes direction when it crosses the interface. A simple explanation of the law of refraction can be obtained by considering two media whose interface is a plane in which a planar wave front travels. If the velocity is v_1 in the first medium and the wave enters the interface at an angle θ_1, while the velocity is v_2 in the second medium and the wave leaves the interface at an angle θ_2, one has

$$\frac{\sin \theta_1}{\sin \theta_2} = \frac{v_1}{v_2}$$

as can be easily seen from Figure 6-1, which uses only the intuitive principle that the wave front remains planar after refraction.

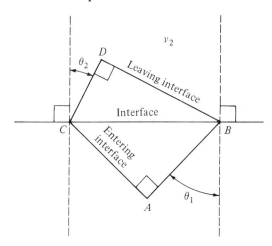

Figure 6-1

For if t is the time required for the ray to traverse AB, then $t = \overline{AB}/v_1$. But obviously $t = \overline{CD}/v_2$, so that $\overline{AB}/\overline{CD} = v_1/v_2$. Then the figure shows that $\overline{CD} = \overline{BC} \sin \theta_2$ and $\overline{AB} = \overline{BC} \sin \theta_1$, which gives the result. If the wave surface is curved, its direction of motion and law of refraction are obtained by considering the tangent plane and using the derivation just given. It is interesting that one can partially check the correctness of the law without even knowing the velocities v_1 and v_2, since the law implies that the ratio $\sin \theta_1/\sin \theta_2$ must remain constant as θ_1 is varied.

One consequence of the law of refraction, if it is correct, is that a wave which is parallel to the interface will not be refracted. For if $\theta_1 = 0$ and θ_2 is not zero, the law of refraction implies $v_1 = 0$. The stubborn inconsistency of this prediction with observation generated a great deal of work.

6.2. Huyghens

In the account of Zigelaar (1981), on which the present section is entirely based, we learn that Huyghens' plan to publish his work on light was held up by one inconvenient fact. His collaborator Jean Picard had obtained a crystal of Iceland spar (calcite) from the Danish Royal Astronomer Rasmus Bartholin, along with Bartholin's little book describing the remarkable refracting properties of this crystal. A ray of light passing through the crystal emerges as two separate beams, only one of which obeys the law of refraction. Five years after he became aware of this fact, Huyghens discovered a possible explanation for it. Along one line in the crystal, called the optical axis, only one velocity is possible for light. In other directions, however, two velocities are possible, one equal to the velocity along the optical axis and the other greater than that velocity. The geometric description of the wave surface of light in such a medium is elegant: It consists of a sphere (for the light which moves with the velocity along the optical axis) plus an ellipsoid of revolution containing the sphere and tangent to it along the optical axis (for propagation with the greater velocity). The spherical nappe of the wave surface has a tangent plane parallel to the interface shown in Figure 6-2 it will therefore not be refracted. The ellipsoidal surface, however, will be refracted.

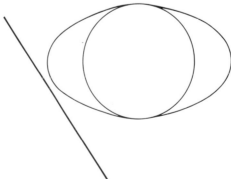

Figure 6-2

6.3.   Fresnel

The phenomenon of double refraction, combined with the double-slit experiments on interference by Thomas Young in the early 1800s, gave rise to new experiments by the French physicists Arago and Fresnel. Two of their memoirs are of particular importance for the problem of the present chapter. In a manuscript dated 30 August 1816 and later found among the papers of Biot, Fresnel wrote (*Oeuvres* II pp. 385–409)

> I had tried in vain to produce fringes using the two images of a luminous point in front of which I had placed a rhomboid of calc-spar . . . I then began to suspect that it was possible that the two systems of waves produced by light in doubly-refracting crystals had no influence on each other. . . .

Fresnel had found, in other words, that the two refracted rays could not be made to interfere with each other, though either could be made to interfere with itself. In short, the two beams were differently polarized.

Fresnel and Arago had been experimenting on polarized light. The fact that the two beams were differently polarized suggested that the velocity of light in such a crystal depended on both its direction of travel and its plane of polarization. In seeking a dynamical basis for the wave theory in terms of undulations of an elastic medium, Fresnel studied crystals and found that some crystals could have two axes of symmetry instead of the single axis of Iceland spar. For such crystals the simple sphere-plus-ellipsoid which Huyghens postulated as the wave surface is not adequate. In a memoir of 31 March 1822 (*Oeuvres* II, pp. 369–442), which was his second supplement to an earlier memoir, Fresnel presented the results of his investigations. Assuming that the wave is a disturbance generated by forces tending to restore a displaced particle, he denoted the components resulting from a unit displacement along the x-axis by a, h, and g; along the y-axis by h, b, and f; and along the z-axis by $g, f,$ and c. (He showed that the components denoted by the same letter here must in fact be the same, using a simple geometrical argument and considering the change in the force of interaction between two molecules when one of them is displaced.) For a displacement making angles X, Y, Z, with the axes the components of the force are then

$$a \cos X + h \cos Y + g \cos Z = p,$$
$$b \cos Y + h \cos X + f \cos Z = q, \qquad (1)$$
$$c \cos Z + g \cos X + f \cos Y = r.$$

After writing these equations Fresnel announced, "I shall now prove that there always exists a direction for which the resultant of these three components is directed along the line of displacement itself, that is, that one can give the angles X, Y, Z real values such that the resultant of these three components shall make angles X, Y, Z with the x-, y-, and z-axes respectively."

The fact Fresnel was claiming will be familiar to anyone who has studied linear algebra. In modern terms it amounts to the statement that a symmetric real matrix always has a real eigenvector (in fact, three mutually perpendicular eigenvectors). Fresnel's argument was based on the following considerations.

The desired relations are $p/r = \cos X/\cos Z$ and $q/r = \cos Y/\cos Z$. Setting $m = \cos X/\cos Z$, $n = \cos Y/\cos Z$ and using system (1), the desired equations become

$$m = \frac{a \cos X + h \cos Y + g \cos Z}{c \cos Z + g \cos X + f \cos Y} = \frac{am + hn + g}{c + gm + fn}$$

with a similar equation for n, namely,

$$n = \frac{bn + hm + f}{c + gm + fn}.$$

The second of these equations is linear in m. Solving it for m and substituting the result in the first equation, one obtains a cubic equation in n, which consequently must have a real solution. Having one such axis, Fresnel immediately showed how to find two more so that the three are mutually perpendicular. He called these directions the *axes of elasticity*.

The axes of elasticity enabled Fresnel to derive the differential equation of the wave surface, namely,

$$\left(z - x\frac{\partial z}{\partial x} - y\frac{\partial z}{\partial y}\right)^2\left(y\frac{\partial z}{\partial x} - x\frac{\partial z}{\partial y}\right) + \frac{\partial z}{\partial x}\frac{\partial z}{\partial y}(a^2 - b^2)\left(z - x\frac{\partial z}{\partial x} - y\frac{\partial z}{\partial y}\right)$$

$$+ \left(a^2 x\frac{\partial z}{\partial y} - b^2 y\frac{\partial z}{\partial x}\right)\left(1 + \left(\frac{\partial z}{\partial x}\right)^2 + \left(\frac{\partial z}{\partial y}\right)^2\right) = 0. \tag{2}$$

Solving this first-order equation certainly does not look trivial. As Fresnel explained:

> The calculations which [this procedure] involved are so long and tedious that I do not think I should record them here. I content myself with saying that I have verified that equation (2) . . . is satisfied by the equation of fourth degree
>
> $$(x^2 + y^2 + z^2)(a^2 x^2 + b^2 y^2 + c^2 z^2) - a^2(b^2 + c^2)x^2$$
> $$- b^2(a^2 + c^2)y^2 - c^2(a^2 + b^2)z^2 + a^2 b^2 c^2 = 0. \tag{3}$$

This neatly symmetric equation for the wave surface was of fundamental importance for the further application of elasticity theory to explain light propagation.

6.4. Lamé

A systematic dynamical theory of elasticity was developed by M. G. Lamé and published in a work entitled *Leçons sur la Théorie Mathématique de l'Élasticité des Corps Solides*. In the second edition (1866) of this work Lamé derived the portions of Fresnel's work just described from elementary principles. His basic equations, derived from Newtonian mechanics, were the conditions for equilibrium:

$$\frac{\partial N_1}{\partial x} + \frac{\partial T_3}{\partial y} + \frac{\partial T_2}{\partial z} + \rho X_0 = 0,$$

$$\frac{\partial T_3}{\partial x} + \frac{\partial N_2}{\partial y} + \frac{\partial T_1}{\partial z} + \rho Y_0 = 0, \qquad (4)$$

$$\frac{\partial T_2}{\partial x} + \frac{\partial T_1}{\partial y} + \frac{\partial N_3}{\partial z} + \rho Z_0 = 0.$$

The meaning of the terms in the first of these equations is as follows. The three components of the force on a plane parallel to the yz-plane through the point (x, y, z) are N_1, T_3, T_2 (the first normal to the plane, the other two tangential to it; ρ is the density of the medium; and X_0 is the x-component of the acceleration of the particle at (x, y, z). Similar considerations apply to the other two equations. The fact that only six letters are needed to denote three components of each of three forces is derived just as Fresnel had done. In Lamé's equations we can explicitly recognize the symmetry.

Lamé showed in a way that will be familiar to students of advanced calculus that in an incompressible medium the displacements (u, v, w) of a particle must satisfy

$$\frac{\partial u}{\partial x} + \frac{\partial v}{\partial y} + \frac{\partial w}{\partial z} = 0. \qquad (5)$$

For the infinitesimal, dx is changed into $(1 + \partial u/\partial x)dx$ by the displacement, so that the volume element $dxdydz$ becomes

$$\left(1 + \frac{\partial u}{\partial x} + \frac{\partial v}{\partial y} + \frac{\partial w}{\partial z}\right)dxdydz,$$

where quadratic terms can be neglected, since all the partial derivatives are believed to be small. Hence if volumes are preserved (the medium is incompressible), equation (5) must hold. Lamé kept the basic functions N_1, N_2, N_3, T_1, T_2, T_3 in a rectangular array which he called a tableau (very close to the notion of matrix)

$$
\begin{array}{ccc}
N_1 , & T_3 , & T_2 \\
T_3 , & N_2 , & T_1 \\
T_2 , & T_1 , & N_3
\end{array}
$$

However, since he had no use for the general operation of matrix multiplication, he was not led to systematize these considerations into linear algebra. His methods were more systematic than those of Fresnel, but still not organized into a theory with propositions and lemmas. From the basic principle that linear magnification is proportional to stress, Lamé showed that each of the six components in the tableau must have the form

$$A\frac{\partial u}{\partial x} + B\frac{\partial v}{\partial y} + C\frac{\partial w}{\partial z} + D\left(\frac{\partial v}{\partial z} + \frac{\partial w}{\partial y}\right) + E\left(\frac{\partial w}{\partial x} + \frac{\partial u}{\partial z}\right) + F\left(\frac{\partial u}{\partial y} + \frac{\partial v}{\partial x}\right).$$

Thus he arrived at a set of 36 functions (6 for each of the 6 components) which characterized the displacements in the medium.

The potential for complicated computation here is enormous. To keep the computations manageable Lamé asked the question (1866, p. 226), "Are these coefficients constant or do they have very short periods? . . . It is impossible to say *a priori*. The comparison of the results given by analysis with those provided by experiment is the only thing that can decide the question." As his contribution toward this effort, he proceeded to obtain the results "given by analysis." He first established conditions necessary for the propagation of a planar wave, where the components of the displacement are given by

$$u = \xi\omega \cos 2\pi\left(\frac{t}{\tau} - \frac{mx + ny + pz}{l}\right)$$

$$v = \eta\omega \cos 2\pi\left(\frac{t}{\tau} - \frac{mx + ny + pz}{l}\right)$$

$$w = \zeta\omega \cos 2\pi\left(\frac{t}{\tau} - \frac{mx + ny + pz}{l}\right)$$

Here ξ, η, ζ are the direction cosines of the displacement, ω is the amplitude of the wave, τ its period, l its wavelength, and m, n, and p the direction cosines of the ray. Equation (5) is satisfied if the displacement is perpendicular to the direction of the ray, i.e., if $m\xi + n\eta + p = 0$.

From the equation for equilibrium, using the principle that a dynamic system reduces to one in equilibrium if its mass times its acceleration is considered a force, we find

$$\frac{\partial N_1}{\partial x} + \frac{\partial T_3}{\partial y} + \frac{\partial T_2}{\partial z} = \rho\frac{\partial^2 u}{\partial t^2} \tag{6}$$

with similar equations for the accelerations of v and w.

For a doubly-refracting medium, Lamé needed two mutually perpendicular directions of displacement (given by ξ, η, ζ) for each direction of travel (given by m, n, p). This requirement leads to substantial symmetry conditions on the 36 coefficients which determine the N's and the T's. The computations are merely high school algebra, but they do occupy several pages. When they have been carried out, equation (6) and its two companion equations become *Lamé's equations:*

$$\frac{\partial^2 u}{\partial t^2} = c^2\frac{\partial}{\partial y}\left(\frac{\partial u}{\partial y} - \frac{\partial v}{\partial x}\right) - b^2\frac{\partial}{\partial z}\left(\frac{\partial w}{\partial x} - \frac{\partial u}{\partial z}\right)$$

$$\frac{\partial^2 v}{\partial t^2} = a^2 \frac{\partial}{\partial z}\left(\frac{\partial v}{\partial z} - \frac{\partial w}{\partial y}\right) - c^2 \frac{\partial}{\partial x}\left(\frac{\partial u}{\partial y} - \frac{\partial v}{\partial x}\right) \tag{7}$$

$$\frac{\partial^2 w}{\partial t^2} = b^2 \frac{\partial}{\partial x}\left(\frac{\partial w}{\partial x} - \frac{\partial u}{\partial z}\right) - a^2 \frac{\partial}{\partial y}\left(\frac{\partial v}{\partial z} - \frac{\partial w}{\partial y}\right)$$

[The system (7) and equation (5) are written in modern notation in Section 5.4.] Lamé made good use of the symmetry of these equations in expressing the direction cosines of the ray in terms of the two possible velocities in that direction:

$$m^2 = \frac{(a^2 - V_1^2)(a^2 - V_2^2)}{(a^2 - b^2)(a^2 - c^2)}$$

$$n^2 = \frac{(V_1 - b^2)(b^2 - V_2^2)}{(a^2 - b^2)(b^2 - c^2)}$$

$$p^2 = \frac{(V_1^2 - c^2)(V_2^2 - c^2)}{(a^2 - c^2)(b^2 - c^2)}$$

Through strenuous combinatorial reasoning he also obtained the equation of the wave surface (3) given by Fresnel.

The hardest part of Lamé's application of elasticity to explain light propagation was his attempt to describe light radiating from a point source. For such radiation the wave surface spreads out homothetically, and the position of the surface after time λ is given by

$$q\lambda^4 - Q\lambda^2 + RP = 0 \tag{8}$$

where

$$q = a^2 b^2 c^2,$$

$$Q = a^2(b^2 + c^2)x^2 + b^2(a^2 + c^2)y^2 + c^2(a^2 + b^2)z^2,$$

$$R = x^2 + y^2 + z^2,$$

and

$$P = a^2 x^2 + b^2 y^2 + c^2 z^2.$$

This fourth-degree equation, for fixed λ, represents a two-napped surface, the two nappes being stuck together at four points which are the points of intersection with two lines through the point of origin of the wave. (The two lines are of course the optical axes.)

From the physical interpretation of the equation, the problem is as follows: Assume the origin undergoes a vibration $u_0 = X_0 \cos 2\pi(t/\tau)$, $v_0 = Y_0 \cos 2\pi(t/\tau)$, $w_0 = Z_0 \cos 2\pi(t/\tau)$. This disturbance will propagate with two different velocities except in the directions of the optical axes, and so will reach the point (x, y, z) at two different times λ_1 and λ_2. From equation (8) we have

$$\lambda_1 = \frac{\sqrt{Q - \sqrt{Q^2 - 4qRP}}}{\sqrt{2q}}.$$

The expression for λ_2 is similar, except that the first negative sign in the numerator becomes a positive sign. For the components of the displacement Lamé assumed

$$u = X_1 \cos 2\pi \frac{t - \lambda_1}{\tau} + X_2 \cos 2\pi \frac{t - \lambda_2}{\tau}$$

with similar expressions for v and w. Here $X_1, X_2, Y_1, Y_2, Z_1, Z_2$ are functions of (x, y, z) to be determined by the requirement that the displacement satisfy Lamé's partial differential equations (7).

Finding this solution took all of lectures 22 and 23 of Lamé's *Leçons*. Though he did not obtain explicitly the form of X_1, X_2, etc., he did find that the amplitude of the vibration was given by

$$\frac{2\epsilon}{b\sqrt{(a^2 - c^2)}\sin i \sin i'} \cdot \frac{1}{\sqrt{x^2 + y^2 + z^2}}$$

where i and i' are the angles between the ray and the two optical axes. Having obtained this solution, Lamé pointed out that the amplitude just given becomes infinite along both optical axes. The 24th and last of his *Leçons* was devoted to explaining away this paradox.

Lamé had pointed out at the end of the 23rd lecture that the components of the displacement are a sum of two terms. Now the sum of two *oppositely* infinite terms may have a finite value; the form $\infty - \infty$ is indeterminate. It is precisely along the optical axes that the magnitudes of X_1 and X_2 both become infinite. Lamé claimed that the indeterminacy could be removed by considering the projections of the path of vibration around the optical axes on both nappes of the surface. He did not give any details, however, and the stubborn fact remained that the amplitude definitely becomes infinite at the origin of the disturbance, due to the factor $(x^2 + y^2 + z^2)^{-1/2}$. To handle this problem Lamé invoked the ether. As he put it,

> Since matter is not continuous, if one partitions the space occupied by the doubly-refracting body into congruent polyhedra, each of which contains a single molecule, every elementary polyhedron will constitute what may be called the system of one molecule. If this is assumed, the origin O is occupied by a molecule having weight, and it is necessary to discover what kind of agitation can take place in its system in order that the result be the progressive waves whose effects are authenticated by experiment . . . But . . . if the vibrating medium which conducts the light in the crystal was made up only of particles having weight, one is ineluctably led to the conclusion that the molecule O must undergo vibration of infinite magnitude in all directions simultaneously. It thus follows necessarily that the central system, and then all the doubly-refracting space must contain another type of matter which is the actual medium vibrating under the influence of the light. Thus the matter having weight plays only a passive role, modifying by a sort of resistance, the directions of the vibrations and the velocities of propagation . . .

This contortion clearly would not do. In the first place it directly contradicted Lamé's derivation of his equations, since they were based on

Newton's laws. One cannot do this and then suddenly assume that in fact the equations are describing the motion of something entirely different from the object they were intended to describe. In addition, whether matter is "ponderable" (Lamé's word) or not, the amplitude Lamé derived remains infinite at the origin, and it is not clear that all the difficulties this fact entails are removed by transferring the condition to some kind of weightless matter. Lamé had been driven to these desperate measures by his desire to retain the solution to system (7) which he had found with so much labor. Others who read his work, in particular Weierstrass, saw another way out of the difficulty: accept the fact that the assumed form of the solution is wrong, and look for other solutions of (7).

6.5.   Weierstrass

Though generally reputed to have been a pure mathematician, Weierstrass often lectured on mathematical physics. In 1856 he even delivered a paper on refraction (*Werke* III, pp. 175–178) at the Congress of German Scientists in Vienna. In this paper he gave a geometric construction for the path of a light ray through several contiguous media where all the refracting surfaces are spherical. Also, as the letter from him to Kovalevskaya on 1 January 1875 (see Section 5.1) shows, he was interested in elasticity theory. His writings, however, leave the impression that he was interested in physics mostly as a source of illustrations of already developed theories. This impression, if correct, is ironic, considering that he decried this sort of mathematics in his *Antrittsrede* at the University of Berlin (*Werke* I, p. 225). His importance in the present chapter is that he posed a problem about light refraction to Kovalevskaya and gave her an unpublished method of his own which he thought would contribute to its solution. The nature of his interest in the problem and his motivation for developing the method are unknown. We are unable to say whether Weierstrass developed the method after studying the elastic solid theory of light propagation or whether he developed it in some other connection and was interested in the equations Lamé had derived because his method seemed applicable to them. All that is certain is that Weierstrass discovered the method around 1861 (cf. *Werke* I, p. 296), that he never published it, and that in 1881 he turned it over to Kovalevskaya to be used as part of her *Habilitationsschrift*. The method is not taught in the usual course of differential equations nowadays; for that reason a discussion in modern notation has been put into Appendix 3 to satisfy the curiosity of mathematicians. The discussion which follows below is intended to be historical and preserves Weierstrass' notation. The only version of the paper we have is the one published by Kovalevskaya as part of her paper (1885a) and reproduced in Weierstrass' collected works (*Werke* I, pp. 275–296).

The central concept in Weierstrass' method is a surface S in three-dimensional space which intersects every ray from the origin in precisely one point. Thus for each point P in space different from 0, there is a unique point

P_1 on the intersection of the ray OP and the surface S. If we set $t = OP/OP_1$, then t is a continuous positive-valued function of P, and the locus $t = $ const. is a surface σ_t similar to S and inside S for $t < 1$, outside S for $t > 1$. The equation of the surface σ_t is

$$u'U + v'V + w'W = t,$$

where (U, V, W) are the coordinates of a point of σ_t and $u' = \partial t/\partial u$, $v' = \partial t/\partial v$, and $w' = \partial t/\partial w$ [evaluated at (U, V, W)].

Denoting by

$$\pm \int_{(t_0 \cdots t)} F(u, v, w)d\omega$$

the integral over the solid region between σ_{t_0} and σ_t (the ambiguous sign being the sign of $t - t_0$), Weierstrass used a "polar-coordinate" formula with σ_t in place of the sphere together with the divergence theorem to derive the formula

$$D_t \int_{(t_0 \cdots t)} D_u F(u, v, w) \, d\omega = D_t^2 \int_{(t_0 \cdots t)} u'F(u, v, w) \, d\omega. \qquad (9)$$

To show the usefulness of this formula Weierstrass applied it to solve the differential equation

$$\frac{\partial^2 \psi}{\partial t^2} = A\frac{\partial^2 \psi}{\partial x^2} + B\frac{\partial^2 \psi}{\partial y^2} + C\frac{\partial^2 \psi}{\partial z^2} + 2A'\frac{\partial^2 \psi}{\partial y \, \partial z} + 2B'\frac{\partial^2 \psi}{\partial z \, \partial x} + 2C'\frac{\partial^2 \psi}{\partial x \, \partial y}$$

under the hypothesis that the function

$$Ax^2 + By^2 + Cz^2 + 2A'yz + 2B'zx + 2C'xy$$

is positive for all real values of x, y, z except $x = y = 0$.

As in the work by Fresnel and Lamé, it is easy to see concepts of linear algebra floating around here, but it is difficult to say how systematic Weierstrass' knowledge of the subject was. Here he seems to be using an ad hoc method, but it may have been so well known that it amounted to a systematic method. Hawkins (1977) has persuasively argued that Weierstrass was developing matrix theory from 1858 on. In the important special case where $A = B = C = a^2$ and $A' = B' = C' = 0$ this equation becomes the wave equation

$$\frac{\partial^2 \psi}{\partial t^2} = a^2 \Delta \psi.$$

Weierstrass pointed out that his method leads to results which had earlier been obtained in the general case of the equation above by Cauchy, and in the special case of the wave equation by Poisson. To solve the equation Weierstrass used as an example one takes the surface S as the ellipsoid whose equation is

$$au^2 + bv^2 + cw^2 + 2a'vw + 2b'wu + 2c'uv = 1$$

where

$$a = \frac{BC - AA'}{G}, \qquad b = \frac{CA - BB'}{G}, \qquad c = \frac{AB - CC'}{G}, \qquad \text{etc.}$$

and

$$G = ABC - AA'A' - BB'B' - CC'C' + 2A'B'C'.$$

Once again, it is hard not to notice that the matrix

$$\begin{pmatrix} a & c' & b' \\ c' & b & a' \\ b' & a' & c \end{pmatrix}$$

is the inverse of the matrix

$$\begin{pmatrix} A & C' & B' \\ C' & B & A' \\ B' & A' & C \end{pmatrix}$$

but Weierstrass consistently wrote out the formulas in full.

Assuming that the solution of the equation has the form

$$\psi(u, v, s; t) = D_t \left(\int_{J(t_0 \cdots t)} \phi(x, y, z) f(u + x, y + v, z + w) \, d\omega \right),$$

where ϕ and f are to be determined, one finds by repeated application of formula (8) that the second-order differential expression in ψ which occurs in the equation is given by

$$D_t^3 \left(\int_{J(t_0 \cdots t)} P(x, y, z) f(u + x, y + v, z + w) \, d\omega \right)$$

$$- D_t^2 \left(\int_{J(t_0 \cdots t)} Q(x, y, z) f(u + x, y + v, z + w) \, d\omega \right)$$

$$+ D_t \left(\int_{J(t_0 \cdots t)} R(x, y, z) f(u + x, y + v, z + w) \, d\omega \right),$$

where P, Q, R can easily be computed in terms of ϕ. If we then choose

$$\phi(x, y, z) = (ax^2 + by^2 + cz^2 + 2a'yz + 2b'zx + 2c'xy)^{-1},$$

computation reveals that $P = \phi$ and $Q = 0 = R$. Hence ψ, defined as above, satisfies the differential equation and f can still be arbitrarily chosen. By suitable choices of f and superposition of solutions (the equation is linear) one can satisfy prescribed initial conditions

$$\psi(u, v, w; 0) = g(u, v, w),$$

$$\frac{\partial \psi}{\partial t}(u, x, w; 0) = h(u, v, w).$$

Weierstrass went on to show that this method, indeed the very solutions just obtained, could be applied to solve much more general equations. To be specific he solved a system of three equations in three unknown functions, of which the first is

$$\frac{\partial^2 \xi}{\partial t^2} + a^2 \left(\frac{\partial}{\partial y} \left(\frac{\partial \eta}{\partial x} - \frac{\partial \xi}{\partial y} \right) - \frac{\partial}{\partial z} \left(\frac{\partial \xi}{\partial z} - \frac{\partial \zeta}{\partial x} \right) \right)$$

$$- b^2 \left(\frac{\partial^2 \xi}{\partial x^2} + \frac{\partial^2 \eta}{\partial x \partial y} + \frac{\partial^2 \zeta}{\partial x \partial z} \right) = X.$$

The other two equations are similar. If we write $\boldsymbol{\xi} = (\xi, \eta, \zeta)$ and $\mathbf{X} = (X, Y, Z)$, the system would be written in modern notation as

$$\frac{\partial^2 \boldsymbol{\xi}}{\partial t^2} + a^2 \operatorname{curl}(\operatorname{curl} \boldsymbol{\xi}) - b^2 \operatorname{grad}(\operatorname{div} \boldsymbol{\xi}) = \mathbf{X}.$$

Weierstrass was able to solve this equation with prescribed initial values of the unknown functions and their first partial derivatives with respect to t. The method is a straightforward application of the results already obtained using an identity which in modern notation would be written

$$\left(\frac{\partial^2}{\partial t^2} - a^2 \Delta \right) \left(\frac{\partial^2}{\partial t^2} - b^2 \Delta \right) = \left(\frac{\partial^2}{\partial t^2} + a^2 \operatorname{curlcurl} - b^2 \operatorname{graddiv} \right)$$

$$\left(\frac{\partial^2}{\partial t^2} + b^2 \operatorname{curlcurl} - a^2 \operatorname{graddiv} \right),$$

where Δ is the Laplacian.

This extension brought Weierstrass to equations very similar, outwardly at least, to those of Lamé. In fact the special case $a = b = c$ in Lamé's equations coincides with the special case $b = 0$, $X = Y = Z = 0$ in the equations above. In addition there is an intuitive link, since Lamé's equations describe a phenomenon involving the wave surface and Weierstrass' solutions rely on a surface for their construction. This similarity gives strong reasons for investigating the possibility of applying Weierstrass' method to Lamé's equations taking the wave surface as the surface S. The only difficulty is that the wave surface lacks the fundamental property on which the method relies: It intersects rays from the origin twice rather than once. On the other hand, each nappe of the wave surface has this property, so that if only some formula could be found which involves only one of the nappes, there might be some hope. The project certainly looked feasible, and it is this problem which Weierstrass posed to Kovalevskaya in mid-1881: Find the general solution of Lamé's equations using the method described above. Kovalevskaya did the work during the two worst years of her life, and as we have seen, finally in 1883 produced a paper which satisfied Weierstrass. It is this work which we now examine.

6.6.   Kovalevskaya

The first half of Kovalevskaya's paper (1885a) is the exposition of Weierstrass' paper summarized above and is enclosed in quotation marks. The second half of the paper contains the application of the method to the system (7).

Following Weierstrass' method, Kovalevskaya assumed solutions to (7) in the form

$$\xi = D_t\left(\int_{(t_0\cdots t)} \phi_1(u, v, w)f(x + u, y + v, z + w)\, d\omega\right),$$

$$\eta = D_t\left(\int_{(t_0\cdots t)} \phi_2(u, v, w)f(x + u, y + v, z + w)\, d\omega\right), \qquad (10)$$

$$\zeta = D_t\left(\int_{(t_0\cdots t)} \phi_3(u, v, w)f(x + u, y + v, z + w)\, d\omega\right).$$

One application of Weierstrass' fundamental formula (9) leads to a system of three equations of which the first is

$$\frac{\partial\xi}{\partial y} - \frac{\partial\eta}{\partial x} = D_t^2\left(\int\left(\phi_1\frac{\partial\theta}{\partial v} - \phi_2\frac{\partial\theta}{\partial u}\right)f(x + u, y + v, z + w)\, d\omega\right)$$

$$- D_t\left(\int\left(\frac{\partial\phi_1}{\partial v} - \frac{\partial\phi_2}{\partial u}\right)f(x + u, y + v, z + w)\, d\omega\right)$$

where from now on $(t_0\cdots t)$ will be omitted. The function θ is defined by $\theta(x, y, z) = t$ when (x, y, z) belongs to σ_t. Nowadays the three equations derived from (10) using (9) would be written (with the same conventions as above for the vector notation)

$$\text{curl } \boldsymbol{\xi} = D_t^2\left(\int((\text{grad }\theta) \times \boldsymbol{\phi})f(x + u, y + v, z + w)\, d\omega\right)$$

$$- D_t\left(\int \text{curl } \boldsymbol{\phi}f(x + u, y + v, z + w)\, d\omega\right).$$

In seeking to satisfy the second-order system (7) one naturally wants the first-order derivatives in such equations as this to drop out. Therefore Kovalevskaya assumed $\partial\phi_1/\partial v - \partial\phi_2/\partial u = 0$, etc. Kovalevskaya pointed out that this equation is satisfied if there is a function ϕ such that $\phi_1 = \partial\phi/\partial u$, $\phi_2 = \partial\phi/\partial v$, and $\phi_3 = \partial\phi/\partial w$. (This fact can be phrased by saying that the curl of a gradient is zero; it amounts to the fact that the mixed second-order partial derivatives are equal.) Next, defining three new functions ψ_1, ψ_2, ψ_3 by the equations

$$\psi_1 = \frac{\partial\phi}{\partial v}\frac{\partial\theta}{\partial w} - \frac{\partial\phi}{\partial w}\frac{\partial\theta}{\partial v}, \qquad \text{etc.}$$

(what would nowadays be written $\boldsymbol{\psi} = \text{grad } \phi \times \text{grad } \theta$), and setting

$$\xi_1 = D_t^2 \left(\psi_1(u, v, w) f(x + u, y + v, z + w) \, d\omega \right), \qquad \text{etc.}$$

she found that Lamé's equations reduced to

$$\frac{\partial^2 \xi}{\partial t^2} = c^2 \frac{\partial \zeta_1}{\partial y} - b^2 \frac{\partial \eta_1}{\partial z},$$

$$\frac{\partial^2 \eta}{\partial t^2} = a^2 \frac{\partial \xi_1}{\partial z} - c^2 \frac{\partial \zeta_1}{\partial x},$$

$$\frac{\partial^2 \zeta}{\partial t^2} = b^2 \frac{\partial \eta_1}{\partial x} - a^2 \frac{\partial \xi_1}{\partial y}.$$

A second application of formula (9) yields another set of three equations, of which the first is

$$c^2 \frac{\partial \zeta_1}{\partial y} - b^2 \frac{\partial \eta_1}{\partial z} = D_t^3 \left(\int \left(c^2 \psi_3 \frac{\partial \theta}{\partial v} - b^2 \psi_2 \frac{\partial \theta}{\partial w} \right) f(x + u, y + v, z + w) \, d\omega \right)$$

$$- D_t \left(\int \left(c^2 \frac{\partial \psi_3}{\partial v} - b^2 \frac{\partial \psi_2}{\partial w} \right) f(x + u, y + v, z + w) \, d\omega \right).$$

Once again to get the system (7) satisfied it is necessary to make the second term in the last formula vanish, i.e., to assume that

$$c^2 \frac{\partial \psi_3}{\partial v} - b^2 \frac{\partial \psi_2}{\partial w} = 0, \qquad \text{etc.}$$

[As before, in modern language this assumption says that $\text{curl}(T\boldsymbol{\psi}) = \mathbf{0}$, where $T(X, Y, Z) = (a^2 X, b^2 Y, c^2 Z)$.] This condition is fulfilled if there is a function ψ such that $a^2 \psi_1 = \partial \psi / \partial x$, etc. With all these reductions Kovalevskaya wound up looking for three functions ϕ, θ, and ψ satisfying the six equations

$$\frac{\partial \phi}{\partial v} \frac{\partial \theta}{\partial w} - \frac{\partial \phi}{\partial w} \frac{\partial \theta}{\partial v} = \frac{1}{a^2} \frac{\partial \psi}{\partial u},$$

$$\frac{\partial \phi}{\partial w} \frac{\partial \theta}{\partial u} - \frac{\partial \phi}{\partial u} \frac{\partial \theta}{\partial w} = \frac{1}{b^2} \frac{\partial \psi}{\partial v},$$

$$\frac{\partial \phi}{\partial u} \frac{\partial \theta}{\partial v} - \frac{\partial \phi}{\partial v} \frac{\partial \theta}{\partial u} = \frac{1}{c^2} \frac{\partial \psi}{\partial w},$$

$$\frac{\partial \psi}{\partial v} \frac{\partial \theta}{\partial w} - \frac{\partial \psi}{\partial w} \frac{\partial \theta}{\partial v} = -\frac{\partial \phi}{\partial u},$$

$$(11)$$

$$\frac{\partial \psi}{\partial w}\frac{\partial \theta}{\partial u} - \frac{\partial \psi}{\partial u}\frac{\partial \theta}{\partial w} = -\frac{\partial \phi}{\partial v},$$

$$\frac{\partial \psi}{\partial u}\frac{\partial \theta}{\partial v} - \frac{\partial \psi}{\partial v}\frac{\partial \theta}{\partial u} = -\frac{\partial \phi}{\partial w}.$$

If equations (11) could be solved, the system (7) would hold with ξ, η, ζ defined by (10) provided the equations $\theta(u, v, w) = t$ define surfaces σ_t of the type required for the Weierstrass method to work. This requirement means that each ray from the origin must intersect the surface exactly once, and – to be strictly rigorous in the application of (9) – that the function θ be continuously differentiable. It is usually possible to relax this last restriction, however.

If θ is held fixed, equations (11) form a system of six linear equations in the six first-order partial derivatives of ϕ and ψ. Since the trivial solution $\phi = $ const., $\psi = $ const. is not of any interest, θ must be such that the determinant of this system vanishes. This requirement gives a differential equation for θ, namely,

$$HF - G + 1 = 0, \tag{12}$$

where

$$H = \left(\frac{\partial \theta}{\partial u}\right)^2 + \left(\frac{\partial \theta}{\partial v}\right)^2 + \left(\frac{\partial \theta}{\partial w}\right)^2,$$

$$F = b^2 c^2 \left(\frac{\partial \theta}{\partial u}\right)^2 + c^2 a^2 \left(\frac{\partial \theta}{\partial v}\right)^2 + a^2 b^2 \left(\frac{\partial \theta}{\partial w}\right)^2,$$

$$G = (b^2 + c^2)\left(\frac{\partial \theta}{\partial u}\right)^2 + (c^2 + a^2)\left(\frac{\partial \theta}{\partial v}\right)^2 + (a^2 + b^2)\left(\frac{\partial \theta}{\partial w}\right)^2.$$

Formidable though equation (12) appears, it was an equation which Lamé had looked at in his *Leçons* and proved that it was satisfied by $\theta(u, v, w) = \lambda(u, v, w)$, provided λ satisfies

$$q\lambda^4 - Q\lambda^2 + RP = 0$$

with q, Q, R, and P as above and $x = u$, $y = v$, $z = w$.

In other words, the surface should be taken as the wave surface. Thus, unless one can find a better solution of equation (12), it is necessary to deal with a two-napped surface in this problem. Kovalevskaya elected to deal with the latter difficulty. Her task then became to work backwards through all the systems of equations she had derived to verify that they all work when the wave surface is chosen as the surface S. To deal with the difficulty she resorted to a recent paper of Heinrich Weber (1878) which gave a parametrization of the wave surface, namely,

$$u = \lambda b \, \text{sn} \, u_1 \, \text{dn} \, u_2,$$

$$v = \lambda a \operatorname{cn} u_1 \operatorname{cn} u_2 ,$$

$$w = \lambda a \operatorname{dn} u_1 \operatorname{sn} u_2 .$$

It is interesting that Weber had discovered this parametrization in connection with theta functions of two variables which transform into products of elliptic theta functions via a transformation of degree 2. Kovalevskaya did not comment on this connection of the paper with her work on reduction of Abelian integrals, of course, since it was not relevant to the job at hand.

The functions are the Jacobian elliptic functions. For those with argument u_1 the modulus k is given by

$$k^2 = \frac{a^2 - b^2}{a^2 - c^2} ,$$

while for those with argument u_2 the modulus k is replaced by μ, where

$$\mu^2 = \frac{a^2}{b^2} \frac{b^2 - c^2}{a^2 - c^2} .$$

It is then tedious but not difficult to verify that (u, v, w) traces the outer nappe of the wave surface as (u_1, u_2) traverses the rectangle $-K \leq u_1 \leq K$, $-2L \leq u_2 \leq 2L$, where

$$K = \int_0^1 \frac{du_1}{\sqrt{(1 - u_1^2)(1 - k^2 u_1^2)}} , \qquad L = \int_0^1 \frac{du_2}{\sqrt{(1 - u_2^2)(1 - u^2 u_2^2)}} .$$

Similarly, (u, v, w) traces the inner nappe of the wave surface as (u_1, u_2) traverses the rectangle whose corners are $(K \pm iK_1, L \pm 2L_1)$, where K_1 and L_1 are obtained from K and L by replacing k^2 by $1 - k^2$ and μ^2 by $1 - \mu^2$.

Equations (11) are very symmetrical when translated into Weber coordinates. The computations are again tedious but not difficult. The first and fourth of the equations will serve as a suitable sample. They are

$$\frac{\partial \phi}{\partial u_1} \left(\frac{\partial u_1}{\partial v} \frac{\partial \lambda}{\partial w} - \frac{\partial u_1}{\partial w} \frac{\partial \lambda}{\partial v} \right) + \frac{\partial \phi}{\partial u_2} \left(\frac{\partial u_2}{\partial v} \frac{\partial \lambda}{\partial w} - \frac{\partial u_2}{\partial w} \frac{\partial \lambda}{\partial v} \right) = \frac{1}{a^2} \frac{\partial \psi}{\partial u}$$

and

$$\frac{\partial \psi}{\partial u_1} \left(\frac{\partial u_1}{\partial v} \frac{\partial \lambda}{\partial w} - \frac{\partial u_1}{\partial w} \frac{\partial \lambda}{\partial v} \right) + \frac{\partial \psi}{\partial u_2} \left(\frac{\partial u_2}{\partial v} \frac{\partial \lambda}{\partial w} - \frac{\partial u_2}{\partial w} \frac{\partial \lambda}{\partial v} \right) = - \frac{\partial \phi}{\partial u} .$$

Kovalevskaya easily showed that ϕ and ψ must be linear functions of u_1, u_2 and λ, and she selected the particular solution $\phi = u_2$, $\psi = -bu_1$. Thus her final formula for the solution of the system (7) was

$$\xi(u, v, w) = D_t \left(\int_{(t_0 \cdots t)} \frac{\partial u_2}{\partial x} f(u + x, y + v, z + w) \, d\omega \right),$$

$$\eta(u, v, w) = D_t\left(\int_{(t_0\cdots t)} \frac{\partial u_2}{\partial y} f(u + x, y + v, z + w)\, d\omega\right),$$

$$\zeta(u, v, w) = D_t\left(\int_{(t_0\cdots t)} \frac{\partial u_2}{\partial z} f(u + x, y + v, z + w)\, d\omega\right).$$

To verify the correctness of this solution it is necessary to differentiate under the integral sign, and Kovalevskaya realized that some problems might arise here, since all the partial derivatives of u_2 become infinite along the optical axes. She claimed, however, that it is easy to see that the particular combination of derivatives needed in the system (7) can be obtained correctly by differentiating under the integral sign provided the function $f(x, y, z)$ has continuous partial derivatives in the region of integration. She gave no proof of this claim, and so it is not certain why she believed that the particular combination of derivatives she was dealing with was an exception to the general rule. One might conjecture that the form these integrals assume when the coordinates (x, y, z) are changed to Weber coordinates (u_1, u_2, λ) led her to this conclusion. It is easy to verify that

$$\frac{\partial u_2}{\partial x} = \frac{1}{\lambda V} \frac{k^2 - 1}{b} \operatorname{sn} u_1 \operatorname{sn} u_2 \operatorname{cn} u_2,$$

where

$$V = 1 - k^2 \operatorname{sn}^2 u_1 - \mu^2 \operatorname{sn}^2 u_2 - (1 - k^2 - \mu^2)\operatorname{sn}^2 u_1 \operatorname{sn}^2 u_2.$$

Since $\operatorname{sn}^2 u_1 = \operatorname{sn}^2 u_2$ along the optical axes, we have $V = 0$ there, so that $\partial u_2/\partial x = \infty$ on these lines. But, since the element of volume transforms as $d\omega = dxdydz = a^2 b\lambda^2 Vd\lambda du_1 du_2$, the singularity in the integrand is cancelled by the element of volume. This fact may have led Kovalevskaya to make the claim she made. Unfortunately, in order to carry out the transformation on which formula (9) is based it is necessary to consider not only $\partial u_2/\partial x$ but also some other integrals involving derivatives as well. Thus, for instance, we need

$$\frac{\partial \xi}{\partial u} = D_t^2\left(\int_{(t_0\cdots t)} \frac{\partial u_2}{\partial x} \frac{\partial \lambda}{\partial x} f(u + x, y + v, z + w)\, d\omega\right)$$

$$-D_t^2\left(\int_{(t_0\cdots t)} \frac{\partial^2 u_2}{\partial x^2} f(u + x, y + v, z + w)\, d\omega\right).$$

In the first term here both factors in the integrand contain a factor of $1/V$; thus the singularity of this integral is *not* cancelled by the element of volume.

As it turns out Kovalevskaya's purported solution of (7) is not a solution at all, as could be seen by simply taking $f(x, y, z) = y$. The result is then $\xi = 0$, $\eta = 0$, $\zeta = Ct^2$, where C is a constant. These functions clearly do not satisfy (7), since C is positive.

How did Kovalevskaya, Weierstrass, and Runge (who, it will be re-called, proofread the paper for Kovalevskaya) come to overlook this mistake? Obviously none of the three was deeply interested in the result. In Runge's case the lack of interest needs no explanation; proofreading articles by your friends is seldom a labor of love. As for Weierstrass, we may recall that he was having one of his periodic bouts of lassitude at the time the paper was written (1884). Finally, even the author of the paper must be excused, since she had been assigned the topic purely to establish her reputation as a math-ematician at a time when her personal life was in chaos.

The real source of the difficulty seems to be the nineteenth century way of looking at differential equations, as commented on at the end of Chapter 2. It does no good at all to go on for 30 pages telling your reader how you have obtained the solution to a differential equation. The method by which it was obtained is irrelevant if the solution is correct and still more so if the solution is incorrect. Kovalevskaya was arguing that the solution had to be correct because the operations used to produce it guarantee that it must be. Such an argument can be valid, but it does seem strange that no one bothered to carry out the more reliable verification for even a special case.

Perhaps because the physical theory for which Lamé's equations had been derived (the elastic solid theory of light) was superseded by better theories (electromagnetic theory), Kovalevskaya's paper apparently had very few careful readers. Thus the paper went into the *Acta Mathematica* and remained unchallenged during her lifetime.

CHAPTER 7

The Euler Equations

7.1. Introduction

One of the most important applications of Newtonian mechanics occurs in the study of the motion of a rigid body. To apply Newtonian principles to a system of particles rigidly attached to each other is by no means trivial; and when the mathematization of the problem has been carried out, the differential equations which result are among the most elusive in classical mechanics, almost as interesting as those of the three-body problem. Because of its importance this problem attracted the attention of many great mathematicians. Indeed, anyone who has researched the history of the problem will probably agree that it is harder to find mathematicians who have *not* worked on the problem than to find mathematicians who have. This chapter examines the developments most closely related to the problem as studied by Kovalevskaya, then analyzes her work in detail. Finally, since her work represents the final chapter in the story of closed-form solutions, a comparison will be made with alternative methods, of which the work of Klein is taken as representative.

7.2. Euler

The problem of the motion of a rigid body about a fixed point attracted the attention of Euler at several times during his life. He was not entirely satisfied with his early work (1736), and the formulation of the problem which he eventually settled upon was published in 1758. The principal difficulty lies in the fact that the problem involves the moments of inertia about three mutually perpendicular axes $\int (y^2 + z^2)\, dM$, $\int (z^2 + x^2)\, dM$, and $\int (x^2 + y^2)\, dM$, as well as the three products of inertia $\int yz\, dM$, $\int zx\, dM$, and $\int xy\, dM$. One of the special facts needed in order to obtain a tractable mathematical problem is the fact that three mutually perpendicular axes can be chosen in the body for which the three products of inertia are all zero. Euler says in the introduction to the paper just cited that it was his discovery of the existence of these three principal axes which enabled him to overcome the difficulties of the problem. Another important fact is that the velocity of the individual particles can be decomposed into the sum of a "translational" velocity, which is the same for all particles of the body (and is thus zero when the body has a fixed point), and a "rotational" velocity, which at every point lies in a plane perpendicular to a certain line along which it is zero (this line is called the instantaneous axis of rotation). The rotational velocity is directly proportional to the distance from the particle to the instantaneous axis of rotation. All these facts Euler took as known to the reader of his paper.

To study the general motion of a body when its center of mass is fixed, Euler took as parameters the angular velocity ४ and the three angles α, β, γ which the instantaneous axis of rotation makes with three mutually perpendicular axes IA, IB, IC through the center of mass I. (These four parameters are not independent, since $\cos^2 \alpha + \cos^2 \beta + \cos^2 \gamma = 1$.) The problem is to derive differential equations to combine with the obvious equation

$$\cos \alpha \sin \alpha \, d\alpha + \cos \beta \sin \beta \, d\beta + \cos \gamma \sin \gamma \, d\gamma = 0$$

so as to determine α, β, γ, and ४ as functions of time. To do so Euler assumed a particle of mass dM located at a point Z whose coordinates with respect to the three mutually perpendicular axes are x, y, and z. However, because the velocities are difficult to describe in rectangular coordinates, he decided to locate the point by its distance from the center of mass ($s = \sqrt{x^2 + y^2 + z^2}$) and the direction cosines of the line from the center of mass through Z, which he wrote as

$$\cos \widehat{AZ} = \frac{x}{s}, \qquad \cos \widehat{BZ} = \frac{y}{s}, \qquad \cos \widehat{CZ} = \frac{z}{s}.$$

If O is the point on the axis of rotation nearest Z on the sphere of radius s about I, Euler constructed an arc $\widehat{ZR}$ $90°$ in length perpendicular to the arc $\widehat{OZ}$, as in Figure 7-1.

Since $\widehat{RZ}$ is an arc of $90°$ and perpendicular to $\widehat{OZ}$, it follows that $\widehat{OZ}$ is the "equator" of the sphere if R is regarded as the "North Pole." Hence $\widehat{OR}$ is also a $90°$ arc. Then by the basic formula of spherical trigonometry

$$\cos \widehat{AR} = \cos \widehat{OR} \cos \widehat{OA} + \sin \widehat{OR} \sin \widehat{OA} \cos(\sphericalangle AOR),$$

one finds

$$\cos \widehat{AR} = \sin \widehat{OA} \cos (\sphericalangle AOR)$$

Since the velocity of the point Z is parallel to the line segment IR and has magnitude ४s, the x-component of the velocity is $u =$ ४$s \sin \widehat{OZ} \cos \widehat{AR}$. If one takes account of orientation, so that a clockwise rotation counts as negative and counterclockwise as positive (when coordinates are chosen in a plane perpendicular to the axis of rotation in such a way that these axes and the upper half of the axis of rotation form a right-handed system) then the arcs and angles should all be regarded as *directed* arcs and angles; we then have

$$\cos \widehat{AR} = \sin \widehat{AO} \cos \sphericalangle AOR = - \sin \widehat{AO} \sin \sphericalangle AOZ.$$

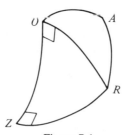

Figure 7-1

Similar equations apply to the other two components v, w, of the velocity. But since $\sphericalangle AOZ = \sphericalangle BAO - \sphericalangle BAZ$ and AB is a 90° arc, one can again apply the formula of spherical trigonometry and obtain finally

$$u = \aleph(z \cos \beta - y \cos \gamma),$$
$$v = \aleph(x \cos \gamma - z \cos \alpha),$$
$$w = \aleph(y \cos \alpha - x \cos \beta).$$

Next, by formal algebraic operations,

$$dx = u\,dt = (z \cos \beta - y \cos \gamma)\,dt, \qquad \text{etc,}$$

leading to the basic equations for acceleration:

$$du = z\,d(\aleph \cos \beta) - y\,d(\aleph \cos \gamma)$$
$$+ \aleph^2(y \cos \alpha \cos \beta + z \cos \alpha \cos \gamma - x \sin^2\alpha)\,dt,$$

$$dv = x\,d(\aleph \cos \gamma) - z\,d(\aleph \cos \alpha)$$
$$+ \aleph^2(z \cos \beta \cos \gamma + x \cos \beta \cos \alpha - y \sin^2\beta)\,dt,$$

$$dw = y\,d(\aleph \cos \alpha) - x\,d(\aleph \cos \beta)$$
$$+ \aleph^2(x \cos \gamma \cos \alpha + y \cos \gamma \cos \beta - z \sin^2 \gamma)\,dt.$$

To use the equations just written, one needs only Newton's law $F = ma$. In applying this principle Euler attempted to use what he called "absolute measures" (units). If the force is p, then the increment of velocity in time dt is $(p/m)\,dt$. Euler assumed p measured "by a weight equal to it." Apparently he wished to be very concrete and specific; if he introduced a physical quantity expressed by a number, he felt obliged to tell how the number was to be obtained. Thus, he said, if g is the height through which a heavy body falls in one second at the location where one has measured that weight (the one equal to the force), the formula $(2gp/m)\,dt$ will lead to correct measures. Actually the physical dimensions in this expression are wrong unless g is regarded as a pure number rather than as a length.

Returning our attention to the particle of mass dM at the point Z, we find that the force acting on it must have components

$$\frac{dM}{2g\,dt}\,du, \qquad \frac{dM}{2g\,dt}\,dv, \qquad \frac{dM}{2g\,dt}\,dw,$$

using Euler's absolute units. It is important to realize that these forces are acting on the particle purely because the body is a rotating rigid body, irrespective of any external forces acting on the body.

Now to describe the motion most simply it is convenient to take the axes IA, IB, IC as the principal axes of the body. As Euler recognized, however, these axes are fixed in the body, while his differential equations for the angles α, β, and γ were derived using axes fixed in space. For that reason he first assumed only that the axes IA, IB, IC were the principal axes at one instant of time. Then introducing the three moments of force (components of the torque, as we now say)

$$P = \frac{dM}{2gdt}(ydw - zdv); \quad Q = \frac{dM}{2gdt}(zdu - xdw); \quad R = \frac{dM}{2gdt}(xdv - ydu),$$

Euler found

$$P = Ma^2\frac{d(\aleph\cos\alpha)}{2gdt} + M(c^2 - b^2)\frac{\aleph^2\cos\beta\cos\gamma}{2g},$$

$$Q = Mb^2\frac{d(\aleph\cos\beta)}{2gdt} + M(a^2 - c^2)\frac{\aleph^2\cos\gamma\cos\alpha}{2g},$$

$$R = Mc\frac{d(\aleph\cos\lambda)}{2gdt} + M(b^2 - a^2)\frac{\aleph^2\cos\alpha\cos\beta}{2g},$$

where

$$a^2 = \int(y^2 + z^2)\,dM, \quad b^2 = \int(z^2 + x^2)\,dM, \quad c^2 = \int(x^2 + y^2)\,dM.$$

Euler pointed out that these equations imply that a body on which no external forces act will not usually rotate at constant angular velocity about a fixed axis. For if α, β, γ, are all constant, the equations become

$$P = M(c^2 - b^2)\frac{\aleph^2\cos\beta\cos\gamma}{2g},$$

$$Q = M(a^2 - c^2)\frac{\aleph^2\cos\gamma\cos\alpha}{2g},$$

$$R = M(b^2 - a^2)\frac{\aleph^2\cos\alpha\cos\beta}{2g}.$$

Hence if a, b, and c are unequal, then P, Q, and R can all vanish only if two of the angles α, β, λ are equal to 90°, i.e., the axis of rotation is one of the principal axes of the body. Euler remarked that it was this question (Can a body on which no forces act rotate about a fixed axis?) which led him to discover the principal axes originally.

From the differential equations just given and the assumption that at a given instant the principal axes are those from which the angles α, β, γ are measured, Euler was able to solve the following problem:

If the body is at rest and subject to arbitrary forces, find the axis IO about which it will begin to turn, and the infinitesimal angular velocity which it will acquire in the element of time dt.

In this case $\aleph = 0$ at the instant in question, so that one has

$$d(\aleph\cos\alpha) = \frac{2gPdt}{Ma^2}, \quad \text{etc.}$$

From this equation and its two companions one can find

$$\cos\beta = \frac{Qa^2}{Pb^2}\cos\alpha, \quad \cos\gamma = \frac{Ra^2}{Pc^2}\cos\alpha;$$

and then the relation $\cos^2\alpha + \cos^2\beta + \cos^2\gamma = 1$ gives

$$\cos\alpha = \frac{P}{a^2\sqrt{P^2/a^4 + Q^2/b^4 + R^2/c^4}}$$

with similar expressions for $\cos\beta$ and $\cos\gamma$, whence it follows that

$$d\mathbf{x} = \frac{2g}{M}\sqrt{\frac{P^2}{a^4} + \frac{Q^2}{b^4} + \frac{R^2}{c^4}}\, dt.$$

From this beginning Euler moved on to the more difficult problems which result from the fact that the principal axes of the body do not remain fixed in space. To do so he introduced a fixed circle on the sphere about I through Z containing two fixed points P and Q. (The points P and Q are not to be confused with the moments of force Euler just finished talking about, which were denoted by the same letters.) If the great circles from A, B, C to P are denoted l, m, and n and the angles QPA, QPB, QPC are denoted λ, μ, ν, the spherical trigonometry formula $\cos\widehat{AB} = \cos l \cos m + \sin l \sin m \cos(\mu - \lambda)$ gives

$$\cos(\mu - \lambda) = -\frac{\cos l \cos m}{\sin l \sin m}.$$

Then, no longer needing the old rectangular coordinates, Euler appropriated the letters used for them and wrote $x = \mathbf{x}\cos\alpha$, $y = \mathbf{x}\cos\beta$, $z = \mathbf{x}\cos\lambda$, the angles again being measured from the principal axes of the body. Thus, besides the four variables introduced at the beginning to describe the motion, Euler now had six additional ones to consider. The number of relations connecting these variables is very large. Euler was able to sort through them and derive a system of differential equations which were reasonably symmetrical, but rather complicated. His description of these equations seems prophetic in the light of the later history of this problem:

> In comparing this method of determining the rotational motion with the attempts which I have set forth previously, one notices immediately some very real advantages, especially in regard to the applications one wishes to examine. And when difficulties are still encountered because of the large number of variables, it is no longer in Mechanics that one should seek the means to overcome them, since it seems that the nature of such a motion is not susceptible of any simpler calculation. Everything then reverts to the calculator, who must render the necessary support in Analysis so as to solve the equations which determine the motion; but it is indubitable that there is an infinity of cases which are absolutely unsolvable due to the limitations of Analysis . . . .

The implied challenge in this last sentence was taken up by many later mathematicians of the first rank. Yet analysis alone did not reach a satisfactory solution of the general case, and the qualitative methods of geometry often proved more useful.

Euler's highest achievement in this paper was the solution of the following problems, now generally known as the Euler case of the motion:

(1)   A solid body not being subject to any force, if it has acquired an arbitrary rotational motion about an axis different from its principal axes, to determine the continuation of this motion:

(2)   A solid body of arbitrary shape not being subject to any forces, if one imparts to it an arbitrary motion, to determine the continuation of that motion.

To solve these problems Euler had derived the following equations:

I. $$dx + \frac{c^2 - b^2}{a^2} \, yz \, dt = 0,$$

II. $$dy + \frac{a^2 - c^2}{b^2} \, zx \, dt = 0,$$

III. $$dz + \frac{b^2 - a^2}{c^2} \, xy \, dt = 0,$$

IV. $d(l \sin l) = (y \cos n - z \cos m) \, dt,$

V. $d(m \sin m) = (z \cos l - x \cos n) \, dt,$

VI. $d(n \sin n) = (z \cos m - y \cos l) \, dt,$

VII. $d(\lambda \sin^2 l) = -(y \cos m + z \cos n) \, dt,$

VIII. $d(\mu \sin^2 m) = -(z \cos n + x \cos l) \, dt,$

IX. $d(\nu \sin^2 n) = -(z \cos l + y \cos n) \, dt.$

Problems (1) and (2) amount to solving these differential equations with prescribed initial values for the variables. In so doing, Euler introduced the angle QPO between the horizontal plane and the plane containing the instantaneous axis of rotation and the line IP, which is fixed in space. The position of the axis of rotation is determined by this angle and the arc $\overset{\frown}{OP}$, which can be calculated from the nine quantities α, β, γ, l, m, n, λ, μ, ν. Euler showed, after pages of combinatorial work involving copious new letters to abbreviate constants, that if units are properly chosen, the cosines of l, m, n are proportional to $a^2 x$, $b^2 y$, and $c^2 z$, and that the angle ϕ is determined by

$$\phi = \int \frac{(H - 2ABCFu) \, du \, \sqrt{G}}{(K - 2ABCGu) \, \sqrt{(\mathscr{A} + 2AU)(\mathscr{B} + 2Bu)(\mathscr{C} + 2Cu)}}$$

where $du = xyz \, dt$, and all other letters represent constants. Finally the arc $\overset{\frown}{OP}$ is determined by

$$\cos \overset{\frown}{OP} = \frac{a^2 x^2 + b^2 y^2 + c^2 z^2}{8 \sqrt{G}}.$$

Having solved the general case, Euler considered the special cases which result when two of the three principal moments of inertia $(a^2, b^2, \text{ and } c^2)$ are equal. He showed that in this case the elliptic integral which gives the angle

ϕ becomes an elementary integral. For the case of complete kinetic symmetry ($a = b = c$) he showed that $\aleph$, α, β, and γ are constants.

Two years after writing the paper just described Euler completed a comprehensive treatise on the dynamics of rigid bodies, which was published in 1765. In this treatise the work of the paper just described was systematized using some principles which were later developed in further detail by Lagrange. For instance, in his reprise of the problem of determining the initial axis of rotation produced by a couple of forces, one of which acts at the center of inertia, Euler showed that this axis will be such as to give the kinetic energy (what he called vis viva) an extreme value, which he claimed would be a minimum.

In summary, Euler had laid the foundations for future investigation by deriving the differential equations which must be solved and by solving them in an important special case. Although the equations were not exactly in the form by which they are now known, the principles for deriving them from Newtonian mechanics, especially the use of the principal axes of the body, were well established.

Before leaving Euler, it is of interest to note that in all the work described above he never once used what are known as the Euler angles. (Klein, in his *Theorie des Kreisels,* refers to them as "die sogenannten Eulerschen Winkel." They were discovered by Euler nevertheless and used as generalized coordinates in his 1748 book *Introductio in Analysin Infinitorum,* pp. 366–369. Two of his angles were the complements of the angles now called Euler angles.)

7.3.   Lagrange

Further progress was made in the study of the motion of a rigid body by J.-L. Lagrange, who gave a new formulation of the principles of mechanics in the *Mécanique Analytique.* In the second (1811–1815) edition of this treatise, second part, section IX, §II, he wrote

> We have just seen, in the preceding paragraph, that whatever motion a rigid body may have, that motion depends on only six variables, of which three refer to the motion of a unique point of the body, which we called the center of the system, while the other three serve to determine the rotation of the body about this center.

Lagrange wrote the components of the translational force on a body as

$$\frac{d^2x'}{dt^2}m + SXDm, \qquad \frac{d^2y'}{dt^2}m + SYDm, \qquad \frac{d^2z'}{dt^2}m + SZDm,$$

where x', y', z' are the rectangular coordinates of the center of the system with respect to axes fixed in space, and X, Y, Z are the components of the force per unit mass acting on the particle of mass Dm. The letter S denotes summation over the particles Dm. For the rotational motion the components of the torque must be

$$S\left(\eta\frac{d^2\zeta}{dt^2} - \zeta\frac{d^2\eta}{dt^2} + \eta Z - \zeta Y\right)Dm,$$

$$S\left(\zeta\frac{d^2\xi}{dt^2} - \xi\frac{d^2\zeta}{dt^2} + \zeta X - \xi Z\right)Dm,$$

$$S\left(\xi\frac{d^2\eta}{dt^2} - \eta\frac{d^2\xi}{dt^2} + \xi Y - \eta X\right)Dm,$$

where ξ, η, ζ are rectangular coordinates of the element Dm with the center of the system as origin and axes parallel to the axes fixed in space.

Lagrange made use of the fact that every point ξ, η, ζ in a rotating body can be expressed as

$$\xi = a\xi' + b\xi'' + c\xi''',$$

$$\eta = a\eta' + b\eta'' + x\eta''',$$

$$\zeta = a\zeta' + b\zeta'' + c\zeta''',$$

where, as he said, a, b, and c depend only on ξ, η, ζ and not on time, while the quantities ξ', ξ'', ξ''', η', η'', η''', ζ', ζ'', ζ''' depend only on time and not on ξ, η, ζ. As we would phrase it, if $\xi = a$, $\eta = b$, $\zeta = c$ at a time t_0, then $[\xi(t),\ \eta(t),\ \zeta(t)]$ is given by $R(t)(a, b, c)$, where $R(t)$ is a rotation whose matrix has the nine entries ξ', ξ'', etc.

Lagrange analyzed the rotation problem by forming the quantities

$$T = S\left\{\frac{1}{2}\left[\left(\frac{dx}{dt}\right)^2 + \left(\frac{dy}{dt}\right)^2 + \left(\frac{dz}{dt}\right)^2\right]\right\}Dm$$

which is the kinetic energy and

$$V = S\left(\int \bar{P}d\bar{p} + \bar{Q}d\bar{q} + \bar{R}d\bar{r} + \cdots\right)Dm$$

which is the potential energy. Here each particle is acted on by the forces $\bar{P}$, $\bar{Q}$, $\bar{R}$, ... proportional to arbitrary functions of the distances $\bar{p}$, $\bar{q}$, $\bar{r}$, ... from the particle to the points where the forces act. The quantities T and V are then to be expressed in terms of variables ξ, ψ, ϕ, etc., and by Lagrange's formulation of mechanics

$$0 = \left(d\frac{\delta T}{\delta d\xi} - \frac{\delta T}{\delta \xi} + \frac{\delta V}{\delta \xi}\right)\delta\xi + \left(d\frac{\delta T}{\delta d\psi} - \frac{\delta T}{\delta \psi} + \frac{\delta V}{\delta \psi}\right)\delta\psi$$

$$+ \left(d\frac{\delta T}{\delta d\phi} - \frac{\delta T}{\delta \phi} + \frac{\delta V}{\delta \phi}\right)\delta\phi + \cdots. \tag{1}$$

In the *Mécanique Analytique*, first part, fourth section, §I, paragraph 10, Lagrange had explained the notation as follows: dx is an infinitely small change in position at x independent of the change in position of other particles in the body. If the entire body undergoes an infinitely small change in position, then the change in the x-coordinate of a particle is denoted δx. The rules for operating with δ are essentially the same as for the operator d. That is, if $U = U(x, y, z)$, then

$$\delta U = \frac{\partial U}{\partial x} \delta x + \frac{\partial U}{\partial y} \delta y + \frac{\partial U}{\partial z} \delta z.$$

For derivatives (e.g., y'), one has

$$\delta y' = \delta\left(\frac{dy}{dx}\right) = \frac{dx\delta(dy) - dy\delta(dx)}{(dx)^2}$$

$$= \frac{\delta dy}{dx} - y'\frac{\delta dx}{dx}$$

$$= \frac{d\delta y}{dx} - y'\frac{d\delta x}{dx}$$

$$= \frac{d(\delta y - y'\delta x)}{dx} + y''\delta x.$$

The important rule $d\delta = \delta d$ was justified according to Lagrange because the two operators are of different types and hence act independently of each other.

Lagrange claimed that when the variables are independent, one has only to set the coefficients of each variation in equations (1) equal to zero to obtain the equations which describe the motion. This statement is essentially the claim that the integral $\int(T - V)\,dt$ will have a minimum, which means (cf. Appendix 5)

$$\frac{d}{dt}\left(\frac{\delta(T - V)}{\delta\xi'}\right) = \frac{\delta(T - V)}{\delta\xi}, \quad \text{etc.}$$

where $\xi' = d\xi/dt$. Since V does not depend on ξ',

$$\frac{\delta(T - V)}{\delta\xi'} = \frac{\delta T}{\delta\xi'}.$$

Then

$$\delta\xi' = \frac{d(\delta\xi - \xi'\delta t)}{dt} + \xi''\delta t = \frac{d\delta\xi}{dt} = \frac{\delta d\xi}{dt}$$

since $\delta t = 0$. Thus we have

$$\frac{d}{dt}\left(\frac{\delta(T - V)}{\delta\xi'}\right) = \frac{d}{dt}\left(\frac{\delta T}{\delta\xi'}\right) = \frac{d}{dt}\left(\frac{\delta T}{\delta d\xi/dt}\right) = d\left(\frac{\delta T}{\delta d\xi}\right).$$

Hence the motion is such that

$$d\left(\frac{\delta T}{\delta d\xi}\right) - \frac{\delta T}{\delta \xi} + \frac{\delta V}{\delta \xi} = 0.$$

For a rotating rigid body Lagrange found

$$T = \tfrac{1}{4}(Ap^2 + Bq^2 + Cr^2) - Fqr - Gpr - Hpq,$$

where A, B, and C are the moments of inertia which Euler had denoted a^2, b^2, c^2, and F, G, H are the products of inertia.

The quantities p, q, r are the components of the angular velocity. Apparently it was Lagrange who first used the Euler angles to express the quantities p, q, and r. (He attributed the use of these angles to d'Alembert, but did not say where d'Alembert had used them.) The Euler angles, denoted ω, ψ, ϕ, are described as follows. One set of axes (xyz) is fixed in space and another set (XYZ) is fixed in the body. The xy-plane and the XY-plane then either coincide, or (what is usually the case) intersect in a line called the line of nodes. The dihedral angle between these two planes (which is also the angle between the z-axis and the Z-axis) is the angle ω. Unless the planes coincide (in which case $\omega = 0°$ or $180°$), the angle from the x-axis to the line of nodes is denoted ψ, and the angle from the line of nodes to the X-axis is denoted ϕ. These three angles are called, respectively, the angle of nutation, the angle of precession, and the angle of (self-)rotation. As mentioned above, the angles Euler actually defined were $\zeta = \tfrac{1}{2}\pi - \phi$, $\eta = \omega$, and $\theta = \tfrac{1}{2}\pi - \psi$. In terms of the Euler angles the components p, q, and r are expressed as

$$p = \frac{\sin \phi \sin \omega \, d\psi + \cos \phi \, d\omega}{dt}, \qquad q = \frac{\cos \phi \sin \omega \, d\psi - \sin \phi \, d\omega}{dt},$$

$$r = \frac{d\phi + \cos \omega \, d\psi}{dt}.$$

Lagrange did not find any expression for V or $\delta V/\delta \omega$, etc., since, he said, there is no difficulty in doing so. He did give a long discussion of the case in which the center of gravity of the body is fixed, showing that his equations for the motion in this case coincide with those derived by Euler. He then turned to the more general case of rotation about a point different from the center of gravity (second part, section IX, § III, paragraph 34). By taking one coordinate axis through the center of gravity and taking the origin at the fixed point, Lagrange obtained the equations of motion in the following form:

$$\frac{d}{dt}\left(\frac{dT}{dp}\right) + q\frac{dT}{dr} - r\frac{dT}{dq} + km\zeta'' = 0,$$

$$\frac{d}{dt}\left(\frac{dT}{dq}\right) + r\frac{dT}{dp} - p\frac{dT}{dr} - km\zeta' = 0,$$

$$\frac{d}{dt}\left(\frac{dT}{dr}\right) + p\frac{dT}{dq} - q\frac{dT}{dp} = 0,$$

where k is the distance from the fixed point to the center of gravity, m is the mass of the body, and ζ', ζ'', ζ''' are the coordinates (with respect to axes fixed in the body) of a point one unit directly below the fixed point. Since the components of the gravitational force are then $(m\zeta', m\zeta'', m\zeta''')$, and the coordinates of the center of gravity are $(0, 0, k)$, the components of the torque on the body will be $(-km\zeta'', km\zeta', 0)$, which explains the final members of each of these equations.

In the eighteenth century solving a system of three equations like the one just written meant finding independent integrals, i.e., functions of p, q, r, ζ', ζ'', ζ''' which are constant when these variables satisfy the differential equations. Lagrange immediately found two integrals, namely,

$$f = p\frac{dT}{dp} + q\frac{dT}{dq} + r\frac{dT}{dr} - T - k\zeta''',$$

$$h = \zeta'\frac{dT}{dp} + \zeta''\frac{dT}{dq} + \zeta'''\frac{dT}{dr}$$

He then commented:

> It seems difficult to find any other integrals and consequently to solve the problem in general. However, one can do so provided the shape of the body satisfies special conditions. Thus supposing $F = 0$, $G = 0$, $H = 0$, and in addition $A = B$, one will have $dT/dp = Ap$, $dT/dq = Aq$, and the third equation will become $dT/dr = 0$, whose integral is $T = $ const.

The case $F = 0$, $G = 0$, $H = 0$ arises when one uses the principal axes of the body as coordinates, as had been shown by Euler. The other hypotheses stated by Lagrange define what has since been known as the Lagrange case of the motion. It is of the highest importance in applications, since most of the machinery of the world possesses this degree of symmetry (e.g., gyroscopes).

7.4.   Some Nineteenth-Century Work

The work of Euler and Lagrange effectively made the problem of the rotation of a rigid body into a purely mathematical problem and solved this problem in the two situations of most value for practical application. The mathematical content of the equations was by no means exhausted by their work, however, since the general solution of the differential equations was not found. It was inevitable that the new developments in mathematics in the nineteenth century would lead to new attempts to solve these equations in the general case. In addition, the work of Euler and Lagrange would be given fresh formulations. The work which resulted is too varied to be described completely here; this chapter must concentrate on the lines of development which led to

Kovalevskaya's work, in particular the work Kovalevskaya knew best, which was the theory of Abelian integrals. The present section gives a very brief glance at some work along other lines before returning to this theme.

The work of Euler and Lagrange can be described as "analytic" in that the solutions were expressed using formulas which in principle make it possible to calculate numerical values for the variables. Since these formulas involve the integral of the square root of a cubic polynomial, the functions involved are elliptic functions. Such formulas could have only limited usefulness before the work of Abel and Jacobi. There were, however, other ways of studying the motion.

A geometric study of the Euler case was carried out by Poinsot (1834). Poinsot's work is still taught today because of its beauty and simplicity. In essence it says that there is a plane called the invariable plane which is fixed in space. If one considers an ellipsoid whose center is the center of mass of the rotating body and whose three axes lie along the principal axes of the body, each semi-axis being of length equal to the square root of the moment of inertia about that axis, then this ellipsoid (which is rigidly fixed in the body) rolls without slipping on the invariable plane. The motion can then be studied by examining the curves traced out in the invariable plane and on the ellipsoid by the point of contact, which are called, respectively, the herpolhode and the polhode. Other mathematicians who worked on the problem, but whose works cannot be described for lack of space, include Poisson, Cayley, Maxwell, and Sylvester.

The line of development which leads to Kovalevskaya's work is consistent with her primary mathematical interest, Abelian integrals. We have already seen that the solution of the equations in the cases of Euler and Lagrange involve elliptic functions. Jacobi's discovery that the elliptic functions are most simply constructed out of theta functions was therefore certain to affect the approach to the rotation problem. It was Jacobi himself who pointed out this fact, in a work published in both the *Comptes Rendus* and the *Journal für die reine und angewandte Mathematik* in 1849. Using the theta functions

$$\theta(u) = 1 - 2q \cos 2u + 2q^4 \cos 4u - 2q^9 \cos 6u + \cdots$$

$$H(u) = 2q^{1/4} \sin u - 2q^{9/4} \sin 3u + 2q^{25/4} \sin 5u - \cdots$$

and their translates $\theta_1(u) = \theta(K - u)$, $H_1(u) = H(K - u)$, Jacobi found that in the Euler case all nine direction cosines of the body axes in terms of the spatial axes could be elegantly expressed by formulas similar to the following for the angle between the x-axis and the X-axis:

$$\alpha = \frac{\theta_1(0)[H(u + ia) + H(u - ia)]}{2H_1(ia)\ \theta(u)}.$$

Here $u = nt + \tau$, where t is time and a, n, and τ are constants (Jacobi 1849, *Comptes Rendus*, p. 99).

The "analytic" approach was still not entirely successful since the general solution still could not be expressed in closed form. Jacobi's work, however, represented what was hoped to be merely the first step. It gave new hope of finally reducing the general case to some kind of order. As a result the Prussian Academy of Sciences proposed this topic for a competition. The outcome of this competition was announced in the proceedings of the Academy for 5 July 1855 (Leibniz' birthday):

> Although no essay was submitted in the competition for this prize, the physico-mathematical section, having in mind the interest in the subject, decided to repeat the problem in exactly the same wording, as follows: It is known that the number of cases in which the differential equations of analytical dynamics can be integrated in finite form, or even reduced to quadratures, is quite limited; and in view of the repeated efforts which the greatest mathematicians have applied to this subject, it is very probable that most of the mechanical problems whose solution has not yet been achieved in the form mentioned are by their nature not susceptible of integration by quadratures and require the introduction of other analytic forms for their successful handling. Since Jacobi has recently given a beautiful representation in series form of the rotation of a rigid body on which no accelerative force acts, it seems worthwhile to attempt to give a wider expression to the application of series and with their aid to handle cases of rotational motion which have not yet been reduced to quadratures. One such case is offered by the problem of rotation of a heavy body, for which reduction to quadratures has been attained only in one special case due to Lagrange, The Academy therefore makes the complete solution of this problem the subject of a competition and poses the problem:
>
> > "To integrate the differential equations for the motion of a body rotating about a fixed point, on which no accelerative force except gravity acts, by means of regularly progressing series which represent explicitly as functions of time all quantities required for the knowledge of the motion."

The deadline for the renewed competition was 1 March 1858, and the prize of 100 ducats was to be made on Leibniz' birthday, 5 July 1858. Once again, however, no one entered the competition.

It seems to have been during the 1850s that the Euler equations assumed the form which is universally used nowadays. According to Klein (*Theorie des Kreisels* I, p. 142) this form is due to the English mathematician R. B. Hayward (1858):

$$A \frac{dp}{dt} = (B - C)qr + \Lambda,$$

$$B \frac{dq}{dt} = (C - A)rp + M,$$

$$C \frac{dr}{dt} = (A - B)pq + N,$$

where A, B, C, p, q, r have the meanings given them by Lagrange, and Λ, M, N are the components of the torque on the body.

7.5. Weierstrass

Jacobi's approach to the problem was followed by Weierstrass in the course which he gave at Berlin while Kovalevskaya was there. In the notes from this course (*Werke* VI; Chapters 24–29) he considers the Euler and Lagrange cases of the motion and shows how to express the variables which describe the motion as quotients of the Weierstrass sigma functions, which were a generalization of the elliptic theta functions of Jacobi. These sigma functions play the same role in the general theory of doubly-periodic functions that theta functions play in the theory of Jacobian elliptic functions. The work cited was assembled from notes taken in the summer of 1875, a course which bore the same title as the one given in the summer of 1873, when Kovalevskaya was studying with him. Thus we may be sure Kovalevskaya was familiar with this material. As a matter of fact the fifth section of this volume was used by Kovalevskaya in her paper (1890) on the rotation problem and was printed in Weierstrass' collected works from a manuscript furnished by Kovalevskaya.

This work of Weierstrass is noteworthy in one additional respect: It used the notions of inner and outer products of directed line segments. The representation of them given by Weierstrass, though accompanied by explanations which distinguish between the nature of the two objects, was couched in a notation which made them appear rather similar. The inner product of the segments 0ξ and $0\xi'$ in Figure 7-2 was denoted $\overline{0\xi'} \cdot \overline{0\eta}$ and the outer product was denoted $\overline{0\xi'} \cdot \overline{\eta\xi}$.

Weierstrass was careful to say, however, that the outer product should be thought of as a line segment perpendicular to the two factors. As we have seen in the letter Weierstrass wrote to Kovalevskaya on New Year's Day 1875 (Section 5.1), Weierstrass despised quaternions. He evidently became acquainted with the inner and outer product from the 1862 version of Grassmann's *Ausdehnungslehre*.

Weierstrass was apparently encouraged by his success in these special cases to investigate the general case. If the system can be reduced to quadratures, the resulting integrals should be algebraic integrals, theoretically capable of being handled by means of theta functions, since the solution to the Jacobi inversion problem was now essentially complete. From Kovalevskaya's letter of 21 November 1881 (Section 5.4) we know that Weierstrass found that the general solution cannot be expressed in terms of

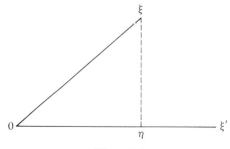

Figure 7-2

single-valued functions of time. He certainly realized that hyperelliptic integrals would be needed, and that the expression of these hyperelliptic integrals in terms of theta functions of several variables would not be simple. Indeed, Weierstrass suggested such a formulation of the problem to Kovalevskaya: Determine the cases in which the functions which describe the motion are Abelian functions of time.

7.6. The Mathematical Mermaid

Before discussing Kovalevskaya's work it is worthwhile to take one closer look at the Euler equations so as to have some idea of the limitations of what can be achieved. For the case when the torque is produced by gravity the form of the equations given by Hayward is

$$A\frac{dp}{dt} = (B - C)qr + Mg(y_0\gamma'' - z_0\gamma'),$$

$$B\frac{dq}{dt} = (C - A)rp + Mg(z_0\gamma - x_0\gamma''), \tag{2}$$

$$C\frac{dr}{dt} = (A - B)pq + Mg(x_0\gamma' - y_0\gamma).$$

where M is the mass of the body, g the acceleration of gravity, $(\gamma, \gamma', \gamma'')$ is a unit vector pointing downward, and (x_0, y_0, z_0) are the coordinates of the center of gravity of the body (with respect to axes fixed in the body). Since the unit vertical vector is fixed in space, we may supplement these equations with three more (cf. Appendix 4):

$$\frac{d\gamma}{dt} = r\gamma' - q\gamma'',$$

$$\frac{d\gamma'}{dt} = p\gamma'' - r\gamma, \tag{3}$$

$$\frac{d\gamma''}{dt} = q\gamma - p\gamma'.$$

Knowing all six of the quantities p, q, r, γ, γ', γ'', we would be able to determine the axis of rotation at any instant and the angular velocity of rotation about that axis, as well as the vertical direction in terms of body coordinates. Now it requires two parameters to determine a direction and one to determine an angular velocity (regarded as a pure number). Hence the solution of these equations will be a five-parameter family of functions. The limitation is not surprising, since the quantities p, q, r, γ, γ', γ'' are not independent, the last three being linked by the fact that the sum of their squares must be unity. On the other hand it is obvious that the physical problem has six degrees of freedom, since the motion is determined by the initial configuration of the body axes in space, for which three parameters are

needed and the initial angular velocity vector, for which three more are
needed. Indeed it is not difficult to see that the functions which figure in these
equations suffice to determine the angles of nutation and self-rotation of the
body, but leave the angle of precession undetermined. The sixth parameter
needed to determine the motion is obtained from the fact that the equations
form an autonomous system, i.e., a system of the form $dW/dt = f(W)$, so
that the right-hand side is independent of time. Hence if $p(t)$, $q(t)$, etc., is a
solution, so is $p(t - t_0)$, $q(t - t_0)$, etc. for any t_0. The arbitrariness of t_0 gives
a sixth degree of freedom to the mathematical problem.

Notice that the Euler equations can be written in the form

$$dt = \frac{dp}{P} = \frac{dq}{Q} = \frac{dr}{R} = \frac{d\gamma}{\Gamma} = \frac{d\gamma'}{\Gamma_1} = \frac{d\gamma''}{\Gamma_2}.$$

If we could solve the last five of these equations, in principle we could then
use the five integrals thereby obtained to express q, r, γ, γ', γ'' in terms of
p. Then P could be expressed in terms of p alone, and the first equation would
give us the complete solution. Thus we are really trying to solve a system of
five equations in six variables. To do so, five independent integrals will be
required. Here again Jacobi contributed some clarifying concepts which,
though not used in Kovalevskaya's work, make it easier to understand where
the difficulty lies. The relevant work is Jacobi's *Werke* IV, pp. 184–255
which describes a method he developed in 1841 which later came to be known
as the theory of multipliers and the method known as the last-multiplier
method (cf. for instance Boole 1849, Chapter 31). The method is explained
in detail in Appendix 6. According to it, a fifth integral can be found provided
one can find four independent integrals and evaluate a certain antiderivative,
which is likely to be possible provided the four integrals found are algebraic
functions. Now it happens that three integrals are easy to find. First, geometry
gives

$$\gamma^2 + (\gamma')^2 + (\gamma'')^2 = 1.$$

Conservation of energy gives the sum of kinetic and potential energies constant, i.e.,

$$\tfrac{1}{2}(Ap^2 + Bq^2 + Cr^2) + Mg(x_0\gamma + y_0\gamma' + z_0\gamma'') = \text{const.}$$

Next, the fact that the force is vertical means that the torque is horizontal;
therefore, the vertical component of the angular momentum is constant, i.e.,

$$Ap\gamma + Bq\gamma' + Cr\gamma'' = \text{const.}$$

At this point we are just one integral away from having a solution of the
general case of the Euler equations when the only force is gravity. Unfortunately this is where we must stop. *The general systems* (2) *and* (3) *have
only three algebraic integrals.* For this reason, as Kovalevskaya remarked,
the Germans called this problem the mathematical mermaid (die mathematische Nixe).

7.7. Kovalevskaya

Since Kovalevskaya's three papers on the rotation problem repeat one another to a large degree, the present exposition will be based on the memoir (1890a) which received the prize from the Academy. The other two papers (1889, 1890b) were written to enlarge on certain points in the basic paper.

Kovalevskaya began by pointing out that in the Euler and Lagrange cases the six unknown quantities p, q, r, γ, γ', γ'' are single-valued functions of time whose only singularities in the finite t-plane are poles. She asked whether the general solution had this property, i.e., whether it had a solution for which all six of these functions could be expressed in a Laurent series about every point. Since the system is autonomous, the point in question may be assumed to be $t = 0$. Thus she asked whether expansions such as the following can be adjusted to meet any possible initial conditions:

$$p = t^{-n_1}(p_0 + p_1 t + p_2 t^2 + \cdots),$$

$$q = t^{-n_2}(q_0 + q_1 t + q_2 t^2 + \cdots),$$

$$r = t^{-n_3}(r_0 + r_1 t + r_2 t^2 + \cdots),$$

$$\gamma = t^{-m_1}(f_0 + f_1 t + f_2 t^2 + \cdots),$$

$$\gamma' = t^{-m_2}(g_0 + g_1 t + g_2 t^2 + \cdots),$$

$$\gamma'' = t^{-m_3}(h_0 + h_1 t + h_2 t^2 + \cdots).$$

By varying the initial conditions at some point $t_0 \neq 0$, we can get five degrees of freedom among the six unknown functions. Hence if such an expansion is always possible, five of the constant coefficients in these expansions must be arbitrary, i.e., not determined by the differential equations (2) and (3). Thus in order to have an affirmative answer to the question we must have two conditions fulfilled:

(1) Each of the six series must have a positive radius of convergence, and the series must formally satisfy the Euler equations.
(2) Five of the coefficients must be left undetermined by the equations.

From the equations it follows easily that $n_1 = n_2 = n_3 = 1$ and $m_1 = m_2 = m_3 = 2$. Then, supposing the unit of length chosen so that $Mg = 1$, we have

$$-Ap_0 = A_1 q_0 r_0 + y_0 h_0 - z_0 g_0,$$

where $A_1 = B - C$, with similar equations for Bq_0 and Cr_0.

In addition $-2f_0 = r_0 g_0 - q_0 h_0$, and again there are similar equations for $-2g_0$ and $-2h_0$. The fact that the leading coefficients are nonzero requires the determinant of this last system of equations to vanish, i.e.,

$$p_0^2 + q_0^2 + r_0^2 + 4 = 0.$$

(It follows from this last result that the leading coefficients cannot all be real numbers. If this result seems strange, remember that the pole was *moved* to zero using the fact that the system is autonomous.) Hence real numbers do not represent time in these equations. In general we would not expect to find a pole at a value of t corresponding to a real value of time.

By combining these equations with the facts that the total energy and the vertical component of the angular momentum are constant, Kovalevskaya worked through the resulting computations and found that unless $A = 2C$ and $x_0 = 0$, the six quantities $p_0, q_0, r_0, f_0, g_0, h_0$ must satisfy one of the following two sets of six equations:

$$p_0 = \epsilon i \frac{2C}{A - 2C} \frac{z_0}{x_0}, \qquad\qquad p_0 = 0,$$

$$q_0 = \epsilon i p_0, \qquad\qquad\qquad q_0 = 2\epsilon i,$$

$$r_0 = 2\epsilon i, \qquad\qquad\qquad r_0 = 0,$$

$$\text{or}$$

$$f_0 = -\frac{2C}{x_0}, \qquad\qquad\qquad f_0 = \frac{2A}{x_0 - i\epsilon z_0},$$

$$g_0 = -i\epsilon \frac{2C}{x_0}, \qquad\qquad\qquad g_0 = 0,$$

$$h_0 = 0, \qquad\qquad\qquad h_0 = i \frac{2A}{x_0 - i\epsilon z_0},$$

where $\epsilon = \pm i$.

Once these initial coefficients are determined, the Euler equations give recursion relations for the six coefficients $p_m, q_m, r_m, f_m, g_m, h_m$, for example,

$$(m - 1)Ap_m - A_1(q_0 r_m + r_0 q_m) + z_0 g_m - y_0 h_m - P_m,$$

where the right-hand side is determined by the coefficients with subscript less than m. Five other equations of similar form hold. Thus there is no option in choosing the coefficients p_0, q_0, etc., unless $A = 2C$ and $x_0 = 0$. If there are to be five arbitrary coefficients in the solution, the various 6-by-6 systems of linear equations which determine the other coefficients recursively must have a total rank deficiency of 5 as m ranges over 1, 2, ... Kovalevskaya said that this requirement meant the determinant must vanish for five positive-integer values of m. Her argument to justify this claim and find out when the conditions were satisfied was given in (1889). Actually this statement is slightly inaccurate. It is possible, though rather unlikely, that the system of equations defining the coefficients with subscript 1, might contain only one independent equation, all the rest being consequences of that one. In such a case we would get five arbitrary constants in the first set of coefficients which we determine after the leading ones, and it would not be necessary for any of the other determinants to vanish. Other variants of the condition are possible as well. In fact, as it turns out, none of these other cases lead to single-valued functions, so that Kovalevskaya's conclusion is correct after all. She may in fact have been through the argument which covers these other cases and merely have forgotten to mention it in her paper. However, this small slip in writing invited criticism, which, as mentioned in Chapter 5, Markov was not slow in providing.

Kovalevskaya ended the first section of her paper by stating, "By carrying out the computations I have verified that these conditions [vanishing of the determinant for five values of m] are not fulfilled in general, but that, besides the two known cases, they are fulfilled also in one new case where the constants satisfy the following equations:

$$A = B = 2C, \quad z_0 = 0.$$

I propose to study this case in the following paragraphs."

By choosing the principal axes in the plane corresponding to A and B suitably, it is possible to arrange that $y_0 = 0$ as well. Then by choosing the unit of length suitably, one can arrange $A = 2 = B$, $C = 1$. Using c_0 to denote Mgx_0, we then have the following system of equations:

$$2\frac{dp}{dt} = qr,$$

$$2\frac{dq}{dt} = -pr - c_0\gamma'',$$

$$\frac{dr}{dt} = c_0\gamma',$$

$$\frac{d\gamma}{dt} = r\gamma' - q\gamma'',$$

$$\frac{d\gamma'}{dt} = p\gamma'' - r\gamma,$$

$$\frac{d\gamma''}{dt} = q\gamma - p\gamma'.$$

The three known integrals can then be written

$$2(p^2 + q^2) + r^2 = 2c_0\gamma + 6l_1,$$

$$2(p\gamma + q\gamma') + r\gamma'' = 2l,$$

$$\gamma^2 + (\gamma')^2 + (\gamma'')^2 = 1.$$

Kovalevskaya quickly found a fourth polynomial integral, which she wrote in complex form as

$$\{(p + qi)^2 + c_0(\gamma + i\gamma')\} \{(p - qi)^2 + c_0(\gamma - i\gamma')\} = k^2.$$

She then began an intricate series of changes of variable in order to transform the integrals and equations to a form suitable for application of theta functions. In this series of transformations the new letters which occur on the left-hand sides of the equations are defined by the first equation in which they occur:

$$x_1 = p + qi,$$

$$x_2 = p - qi,$$

$$y_1 = \gamma + \gamma' i,$$

$$y_2 = \gamma - \gamma' i,$$

$$\xi_1 = x_1^2 + c_0 y_1,$$

$$\xi_2 = x_2^2 + c_0 y_2,$$

$$\mathscr{A} = 6l_1 - (x_1 + x_2)^2,$$

$$\mathscr{B} = 2lc_0 + x_2 x_2 (x_1 + x_2),$$

$$\mathscr{C} = c_0^2 - k^2 - x_1^2 x_2^2,$$

$$\mathscr{R}(x_1) = \mathscr{A} x_1^2 + 2\mathscr{B} x_1 + \mathscr{C},$$

$$\mathscr{R}(x_2) = \mathscr{A} x_2^2 + 2\mathscr{B} x_2 + \mathscr{C},$$

$$\mathscr{R}_1(x_1 x_2) = \mathscr{A}\mathscr{C} - \mathscr{B}^2,$$

$$\mathscr{R}(x_1 x_2) = \mathscr{A} x_1 x_2 + \mathscr{B}(x_1 + x_2) + \mathscr{C},$$

$$W^2 = \{\mathscr{R}_1(x_1 x_2) + k^2(x_1 - x_2)^2\}^2 - 4k^2 \mathscr{R}(x_1)\mathscr{R}(x_2),$$

$$s_1 = \frac{\mathscr{R}(x_1 x_2) - \sqrt{\mathscr{R}(x_1)}\,\sqrt{\mathscr{R}(x_2)}}{2(x_1 - x_2)} + \frac{1}{2} l_1,$$

$$s_2 = \frac{\mathscr{R}(x_1 x_2) + \sqrt{\mathscr{R}(x_1)}\,\sqrt{\mathscr{R}(x_2)}}{s(x_1 - x_2)} + \frac{1}{2} l_1,$$

$$k_1 = \frac{l_1 + k}{2},$$

$$k_2 = \frac{l_1 - k}{2},$$

$$g_2 = k^2 - c_0^2 + 3l_1^2,$$

$$g_3 = l_1(k^2 - c_0^2 - l_1^2) + l^2 c_0^2,$$

$$S = 4s^3 - g_2 s - g_3,$$

$$S_1 = 4s_1^3 - g_2 s_1 - g_3,$$

$$S_2 = 4s_2^2 - g_2 s_2 - g_3,$$

$$\mathscr{R}_1(s) = -S(s - k_1)(s - k_2).$$

Although the reader is not intended to follow the details of all these transformations, it will be well to remark that $\mathscr{R}_1$ is a function of two variables and would nowadays be written with a comma between the variables x_1 and x_2. Also, Kovalevskaya apparently did not notice that she had used the letters g_2 and g_3 in a different sense in the introduction to her paper.

This string of transformations is certainly formidable and gives evidence of either tremendous combinatorial ability or prodigious patience, or both. Of course many of them are merely abbreviations used to keep formulas from becoming too cumbersome. (The formulas become cumbersome, nevertheless!) The transformations arise from the need to evaluate the integrals

$$\int \frac{2}{qr}\,dp = \int dt,$$

after expressing all the other functions in terms of p. (Since it is not feasible to do so, however, the actual procedure circumvents this process.)

Now the variables s_1 and s_2 which occur here are adjusted just so that the procedure which solves the Jacobi inversion problem can be applied. By carefully rewriting the differential equations and the integrals she had found Kovalevskaya was able to show that

$$0 = \frac{ds_1}{\sqrt{\mathcal{R}_1(s_1)}} + \frac{ds_2}{\sqrt{\mathcal{R}_1(s_2)}},$$

$$dt = \frac{s_1\,ds_1}{\sqrt{\mathcal{R}_1(s_1)}} + \frac{s_2\,ds_2}{\sqrt{\mathcal{R}_1(s_2)}},$$

where $\mathcal{R}_1(s)$ is a polynomial of fifth degree. As remarked above, these formulas imply that s_1 and s_2 can be expressed as quotients of products of theta functions whose arguments are linear functions of time. For most mathematicians such a result would be adequate. However, for a student of Weierstrass it is not sufficient to say that you can hear the bird singing in the bush; you have to catch it and display it. Thus, having spent 16 pages obtaining the result that the solutions could *in principle* be expressed in terms of theta functions of two variables, she was now about to spend 50 pages showing how to do so in *practice*.

Although the complete description of the motion in the Kovalevskaya case is eventually going to involve the Jacobi inversion process, the day of reckoning can be postponed, since not all of the variables we need to know are so complicated.

In particular, noting that

$$\mathcal{R}(x_1) = -x_1^4 + 6l_1x_1^2 + 4c_0x_1 + c_0^2 - k^2,$$

and setting

$$u = \int \frac{1}{\sqrt{\mathcal{R}(x)}}\,dx,$$

we can invert this elliptic integral (the exact procedure depending on the location of the roots of the polynomial) and obtain $x = \phi(u)$. By a suitable choice of constants Kovalevskaya found

$$\phi(u) = i\left(h_0 + \frac{2E}{h_1\mathcal{P}u + k_2}\right),$$

where $\mathcal{P}$ is the Weierstrass doubly-periodic function (essentially the second derivative of the sigma function which Weierstrass used to generalize the theta functions of Jacobi); $E = (e_1 - e_2)(e_2 - e_3)(e_3 - e_1)$, and e_1, e_2, e_3 are the roots of $s^2 - g_2 s - g_3 = 0$.

Then with $u_1 = (u + vi)/2$, $u_2 = (u - vi)/2$, which must be the case, since $p + qi = x_1 = \phi(u_1)$ and $p - qi = u_2 = \phi(u_2)$, and p and q are real-valued, we find that $s_1 = \mathscr{P}(u_1 + u_2) = \mathscr{P}(u)$, and $s_2 = \mathscr{P}(u_1 - u_2) = \mathscr{P}(iv)$.

By making use of some combinatorial formulas for the Weierstrassian functions, Kovalevskaya was able to make use of Weierstrass' work (*Werke* VI, Chapters 14–17) to eliminate u_1 and u_2 and obtain expressions for p and q, namely,

$$p = -i\,\frac{L_1 P_1 + M_1 P_2 + N_1 P_3}{LP_1 + MP_2 + NP_3};\qquad q = \frac{E}{LP_1 + MP_2 + NP_3}.$$

In these expressions the constants have the following values:

$$L = i(e_2 - e_3)\sqrt{l_1 + e_1}\,,$$

$$M = i(e_3 - e_1)\sqrt{l_1 + e_2}\,,$$

$$N = i(e_1 - e_2)\sqrt{l_1 + e_3}\,,$$

$$L_1 = (e_2 - e_3)\sqrt{(l_1 + e_2)(l_1 + e_3)}\,,$$

$$M_1 = (e_3 - e_1)\sqrt{(l_1 + e_3)(l_1 + e_1)}\,,$$

$$N_1 = (e_1 - e_2)\sqrt{(l_1 + e_1)(l_1 + e_2)}\,,$$

$$P_\alpha = \sqrt{(s_1 - e_\alpha)(s_2 - e_\alpha)}\,,\qquad \alpha = 1, 2, 3.$$

Thus p and q can be obtained from elliptic functions. The expression for r is obtained from the equation $2(dp/dt) = qr$, and the result is

$$r = -i\,\frac{LP_{23} + MP_{31} + NP_{12}}{LP_1 + MP_2 + NP_3},$$

where

$$P_{23} = \frac{1}{s_1 - s_2}\{\sqrt{(s_1 - e_1)(s_1 - k_1)(s_1 - k_2)(s_2 - e_2)(s_2 - e_3)}$$

$$- \sqrt{(s_1 - e_2)(s_1 - e_3)(s_2 - e_1)(s_2 - k_2)(s_2 - k_2)}\}$$

and similar expressions hold for P_{31} and P_{12}. Kovalevskaya next gave similar formulas for the other three functions γ, γ', γ'' in terms of s_1 and s_2. What remained was the final reckoning with the Jacobi inversion process, so as to express s_1 and s_2 as functions of time using theta functions of two variables. This work is both routine and extremely complicated, since it depends on the location of the zeros of the polynomial $\mathscr{R}_1(s)$. Kovalevskaya's patience (and time also) ran out before all the cases were considered, but she was able to derive formulas for the 6 functions $P_1, P_2, P_3, P_{23}, P_{31}, P_{12}$ which express the original functions. For instance, she showed that

$$P_1 = \frac{\theta_3(v_1, v_2)}{\theta_5(v_1, v_2)},$$

where v_1 and v_2 are linear functions of time. (For the meaning of the sub-scripted theta functions see Chapter 3.)

Kovalevskaya then completed her work by showing that all nine direction cosines of the axes of a coordinate system fixed in space with respect to the principal axes of the body could be expressed as quotients of theta functions of v_1 and v_2. In her later paper she supplemented her results with a description of a physical body satisfying the hypotheses of her case. This model was provided to her by H. A. Schwarz at the request of Weierstrass. It consisted of two right circular cylinders whose axes are parallel, and whose dimensions are adjusted so as to give the correct relations among the three principal moments of inertia and the center of gravity.

The physical model described by Schwarz apparently did not adequately convey the essence of the Kovalevskaya case. A more detailed study of this case was subsequently (1897) carried out by the Russian mathematician N. E. Zhukovsky, who pointed out certain simple aspects of the motion. For instance, the upward-pointing unit vector $-(\gamma, \gamma', \gamma'')$ and the vector $(2p, 2g, 2r)$ which is twice the angular velocity vector have projections on the plane of x and y which remain a constant distance apart (cf. Kovalevskaya's fourth integral for her case of the motion). Despite Zhukovsky's work, however, the Kovalevskaya case is so complicated that no simple description of the motion as a whole is possible.

7.8. Conclusions

Kovalevskaya's paper is assured of a permanent glory because it completes a program implicit in the works of Euler and Lagrange—to solve in an analytic manner the equations of motion. To do so requires cases in which there are enough algebraic integrals to permit a reduction to quadratures, and then a transformation of variables suitable for allowing an application of theta functions. The Kovalevskaya case is the last one in which this program can be carried out, since, due to the work of R. Liouville (1897) it is now known that there are no other cases in which the equations have four independent algebraic integrals. In the direction Kovalevskaya went it is not possible to go further. This point was well stated by Hadamard in the remarks he made at the French Congress of Mathematicians in 1895 (*Oeuvres* IV, p. 1719). The case seems to be of no practical value; and indeed one finds that physics textbooks rarely mention the result and never investigate it nowadays.

But if the analytic approach has reached a dead end, it may yet be that geometry can achieve something. For purposes of comparison it may be of interest to depart from the practice in previous chapters and describe some work on the problem subsequent to that of Kovalevskaya. The most interesting and comprehensive investigations were undertaken by Felix Klein and summarized by him and Sommerfeld in the *Theorie des Kreisels* (four volumes 1897–1910).

Having concentrated on Kovalevskaya, we have discussed mostly the methods of the Berlin analysts up to now. Hence a little space devoted to Klein may also illustrate the contrast between Klein's geometry and Weierstrass' analysis, which to some extent was a contrast between Göttingen and Berlin. The most digestible morsel of Klein's work was described in four lectures which he gave at Princeton University in October 1896 (Klein 1897).

7.9. Klein

Klein's first three lectures were devoted to a reduced version of the Euler case.

Like Weierstrass, Klein used analytic function theory and complex values of time to solve the equations. Unlike him, however, he found a geometrical way to use analytic functions, via conformal mapping. He pointed out that rotations are conformal, orientation-preserving maps of the sphere onto itself. Since the sphere can be mapped conformally onto the extended plane through stereographic projection, it follows that a rotation can be considered as a conformal, orientation-preserving map of the extended complex plane onto itself. Such a mapping is an analytic function, and since it is one-to-one, it is well known that it must have the form

$$f(z) = \frac{\alpha z + \beta}{-\bar{\beta} z + \bar{\alpha}} \quad \text{where} \quad |\alpha|^2 + |\beta|^2 = 1.$$

[The correspondence between pairs (α, β) and rotations is two-to-one, since the pair $(-\alpha, -\beta)$ gives the same rotation as the pair (α, β).] Since Klein wanted parameters which were analytic functions, he actually used transformations

$$f(z) = \frac{\alpha z + \beta}{\gamma z + \delta} \quad \text{with} \quad \alpha \delta - \beta \gamma = 1.$$

The quantities α, β, γ, δ are now known as the Cayley–Klein parameters, although Klein said they had been introduced by Riemann "forty years ago." Evidently he had in mind Riemann's work on the Dirichlet principle (*Werke*, p. 309).

As Klein saw it, stereographic projection treats the two horizontal directions symmetrically and distinguishes the vertical direction, and this is precisely the kind of symmetry found in the problem of a body rotating under the influence of gravity. Therefore these parameters should be the natural ones. Klein took the Euler case and assumed also $A = B = C = 1$. In terms of the Euler angles ω, ψ, ϕ the solution of the equations in this case is known to be

$$t - t_0 = \int_{u_0}^{u} \frac{ds}{\sqrt{P(s)}},$$

$$\phi - \phi_0 = \int_{u_0}^{u} \frac{l - ns}{(1 - s^2)\sqrt{P(s)}}\, ds,$$

$$\psi - \psi_0 = \int_{u_0}^{u} \frac{n - ls}{(1 - s^2)\sqrt{P(s)}}\, ds.$$

where $u = \cos \omega$, and $P(s) = 2ps^3 - 2hs^2 + 2(ln - p)s + 2h - l^2 - n^2$. Since these integrals are elliptic integrals, each of them can be inverted. In particular ω can easily be expressed as a function of time, provided $P(s)$ has three real roots, which is assumed. The advantage of the Cayley–Klein parameters appears in finding ϕ and ψ. The two integrals which define them have singularities at ± 1 on both sheets of the Riemann surface, a total of eight singularities. The observation that

$$\phi - \psi = \phi_0 - \psi_0 + (l - n) \int_{u_0}^{u} \frac{1}{(1 - s)\sqrt{P(s)}}\, ds$$

and

$$\phi + \psi = \phi_0 + \psi_0 + (l + n) \int_{u_0}^{u} \frac{1}{(1 + s)\sqrt{P(s)}}\, ds.$$

leaves only four singularities to deal with. There is no help for the fact that these elliptic integrals are of Legendre's third kind. However, there is a somewhat complicated algorithm for reducing such integrals to a standard form, called the normal form:

$$\int \frac{\sqrt{Q(s)}}{1 - s^2}\, ds \quad \text{where} \quad Q(\pm 1) = 4.$$

The normal form is preferred because, for example, around $s = 1$ the Laurent expansion of the integrand is

$$\frac{1}{s - 1} + c_0 + c_1(s - 1) + \cdots,$$

so that the integral is $\log (s - 1) + f(s)$, where $f(s)$ is analytic at $s = 1$. Similarly, around $s = -1$, one finds the integral equal to $-\log (s + 1) + g(s)$. Taking the exponential of the integral, we find that it is meromorphic and has a simple pole at $s = -1$ and a simple 0 at $s = 1$.

Now if the Cayley–Klein parameters are expressed in terms of the Euler angles, the result is

$$\alpha = (\cos \tfrac{1}{2}\omega)e^{i(\phi + \psi)/2}$$

so that

$$\log \alpha = \log \alpha(u_0) + \frac{1}{2} \log \frac{1}{2} \phi_0 + \psi_0 + \int_{u_0}^{u} \frac{\sqrt{P(s)} + i(n + l)}{2(1 + s)\sqrt{P(s)}} \, ds.$$

Since $P(-1) = -(l + n)^2$, it follows that on one sheet of the Riemann surface the integrand has a simple pole with residue 1 at $s = -1$; on the other sheet it has a removable singularity at $s = -1$. In other words, when the Cayley–Klein parameters are used, the integral automatically assumes normal form. Moreover, each of the four singularities encountered in finding $\phi + \psi$ and $\phi - \psi$ goes to one and only one of the four Cayley–Klein parameters. This elegant result led Klein to exclaim that there must exist a pre-established harmony between these parameters and the physical problem. Because of their simplicity these parameters can be neatly expressed in terms of the Weierstrass sigma functions, for example,

$$\alpha(t) = A e^{\lambda t} \frac{\sigma(t + a)}{\sigma(t)}.$$

Klein used this expression to give a geometric discussion of the motion of the top.

Klein also took a step which Kovalevskaya and Weierstrass had not taken and gave a physical interpretation of complex-variable time. He showed that if one complexifies the Lagrangian and then again restricts time to real values, a complexified dynamical problem can be interpreted as a real dynamical problem with twice the original number of variables. Klein stated that the complexified dynamical problem could be interpreted either as motion in hyperbolic non-Euclidean space or as motion in Euclidean space combined with a strain. He emphasized, however, that these interpretations were merely ways of grouping the phenomena, not to be interpreted as having physical reality.

Both Klein and Kovalevskaya encountered hyperelliptic integrals at the borderline of practicality. For Klein this border was reached in the problem of a free top, which occupied the last of his Princeton lectures. The integrals he encountered were

$$v_1 = \int \frac{ds}{\sqrt{P(s)}(1 + pw - pws^2)}.$$

$$v_2 = \int \frac{s \, ds}{\sqrt{P(s)}(1 + pw - pws^2)}.$$

Klein mentioned that Jacobi had shown how to express these integrals as single-valued analytic functions of theta functions of two variables. Unlike Kovalevskaya, he declined to carry through the details of the process. Klein usually looked for a simpler way out when faced with long computations. In the present instance he turned to one of his favorite subjects, automorphic functions. The difficulty in making the functions single-valued (which led to the use of two-variable functions) was that certain regions in the plane get mapped onto one another with overlap. To overcome this problem Klein looked for a conformal mapping from a polygon all of whose angles are right angles to the region bounded by six intersecting semicircles. By a basic principle known as the Schwarz reflection principle it is very easy to tile the plane with images of the latter region; and the group of reflections involved determines a class of automorphic functions. The tiling itself eliminates the multivaluedness in the original functions. But the constants needed for an explicit construction of this tiling eluded Klein. He invited his listeners and readers to undertake a detailed investigation.

Perhaps the contrast between the analytic and the geometric approaches, as I have called them, is highlighted in Klein's words, taken from the introduction to his Princeton lectures:

> . . . It seems to me better wherever possible to effect a mathematical demonstration by general considerations which bring to light its inner meaning rather than by a detailed reckoning, every step in which the mind may be forced to accept as incontrovertible, and yet have no understanding of its real significance.

From the analysis given above, it is manifest that Kovalevskaya hoped rather to bring the inner meaning to light by being very explicit about the formulas which describe the motion.

7.10. Evaluation

As pointed out in Chapter 5, Kovalevskaya's contemporaries were lavish in their praise of her work on this problem. The reason for this praise is not hard to discern. By the late nineteenth century, analysis had grown far in excess of its applications. As Klein said, again in his introduction to the Princeton lectures:

> . . . If one may accuse mathematicians as a class of ignoring the mathematical problems of the modern physics and astronomy, one may, with no less justice perhaps, accuse physicists and astronomers of ignoring departments of the pure mathematics which have reached a high degree of development and are fitted to render valuable service to physics and astronomy . . .

In this context Kovalevskaya's paper had the merit of actually using large parts of recently developed analysis (theta functions, for instance) to solve an old problem of mechanics. Recall that the Bordin committee had mentioned this aspect of the work in its report (Section 5.13). The same point was made by Weierstrass in a letter of 16 August 1888 to H. A. Schwarz, in which he requested a description of a physical model of the Kovalevskaya case and some formulas on elliptic functions which Kovalevskaya needed for her memoir:

> This problem, together with the very cleverly *contrived* problem of the motion of a point mass free of external forces but confined to a surface of second degree, which C. Neumann took for his doctoral dissertation, is the first significant problem of mechanics to be solved by hyperelliptic functions.

Hermite also praised Kovalevskaya for using theta functions. In his letter recommending her for reappointment at Stockholm in 1889 he said that these functions amounted to a new technique for solving differential equations. Apparently the technique has not been much used. Koblitz (1983, p. 254) points out that the next use of two-variable theta functions occurred in a paper of Dubrovin and others on nonlinear equations of Korteweg–de Vries type (1976). Besides the work of Neumann (1871), which reversed the usual procedure and listed some physical problems which could be solved using hyperelliptic functions, there were a few other occasions on which hyperelliptic integrals had arisen in physics. For example Poisson had exhibited some in his 1833 *Traité de Mécanique*, Vol. 2, p. 733. He did not attempt to evaluate them, however. They also arise, as mentioned in Chapter 3, in finding the capacitance of a condenser consisting of one square prism inside another, only in this case the integrals reduce to elliptic integrals.

A fair summary, I believe, is that Kovalevskaya's work is an ingenious application of mathematics to a system of equations of great mathematical interest. Her result is of permanent interest; but since the case to which it applies is rather special, the details of her arguments are no longer worth troubling about.

CHAPTER 8

Bruns' Theorem

8.1. Introduction

The paper on Bruns' theorem was undoubtedly considered by Kovalevskaya a casual work, the sort of thing one communicates in an offhand manner as a point of mild interest. It was in fact a simplified proof of a lemma needed for Bruns' proof of his theorem.

Bruns, a student of Weierstrass, is best remembered for having proved that the differential equations of the three-body problem have only the 10 algebraic integrals derivable from elementary considerations (Bruns 1887). The theorem whose proof Kovalevskaya simplified was *a* theorem of Bruns. It asserts that if a homogeneous solid body is bounded by a surface $W(x, y, z) = 0$, where W is an analytic function, then at every regular point of the boundary surface (i.e., every point where at least one of the partial derivatives of W is not zero) the potential of the body is an analytic function. (It is clear that the potential will be analytic everywhere except the boundary surface.) Kovalevskaya's paper (1891) gave a simplified proof of the following lemma, used in the proof of Bruns' theorem:

There exists a function $U(x, y, z)$ which is analytic at every regular point of the boundary surface S and satisfies Poisson's equation

$$\Delta U = \frac{\partial^2 U}{\partial x^2} + \frac{\partial^2 U}{\partial y^2} + \frac{\partial^2 U}{\partial z^2} = -4k\pi$$

together with the boundary condition $0 = U = \partial U/\partial x = \partial U/\partial y = \partial U/\partial z$ on S.

8.2. Kovalevskaya's Paper

The proof Kovalevskaya gave for the lemma just stated was straightforward. She changed variables from (x, y, z) to (u, v, s) in such a way that the boundary surface has the equation $s = 0$. Then she applied the Cauchy–Kovalevskaya theorem to infer that there is a unique solution to Poisson's equation in the transformed coordinates (u, v, s) which is analytic at $s = 0$ and satisfies $0 = U(u, v, 0) = \partial U/\partial s (u, v, 0)$. Because of the first of these two equations we also have $\partial U/\partial u = 0$ and $\partial U/\partial v = 0$ on S. Then the chain rule, e.g.,

$$\frac{\partial U}{\partial x} = \frac{\partial U}{\partial u}\frac{\partial u}{\partial x} + \frac{\partial U}{\partial v}\frac{\partial v}{\partial x} + \frac{\partial U}{\partial s}\frac{\partial s}{\partial x}$$

implies that $0 = \partial U/\partial x = \partial U/\partial y = \partial U/\partial z$ on S.

The technical details are also straightforward. To get the new coordinates Kovalevskaya parametrized the surface S locally by any analytic coordinates (u, v). For each point (x_1, y_1, z_1) on S (whose coordinates are then functions of u and v) she defined the transformation

$$x = x_1 + \xi s,$$
$$y = y_1 + \eta s,$$
$$z = z_1 + \zeta s,$$

where (ξ, η, ζ) is the unit normal at (x_1, y_1, z_1), hence also a function of u and v. This set of equations defines x, y, z as functions of u, v, s. In a neighborhood of any regular point these equations are invertible, i.e., the determinant

$$\Omega = \begin{vmatrix} \dfrac{\partial x}{\partial u} & \dfrac{\partial x}{\partial v} & \dfrac{\partial x}{\partial s} \\[2mm] \dfrac{\partial y}{\partial u} & \dfrac{\partial y}{\partial v} & \dfrac{\partial y}{\partial s} \\[2mm] \dfrac{\partial z}{\partial u} & \dfrac{\partial z}{\partial v} & \dfrac{\partial z}{\partial s} \end{vmatrix}$$

does not vanish.

Now Poisson's equation transforms to

$$\Omega \frac{\partial^2 U}{\partial s^2} + A_1 \frac{\partial^2 U}{\partial u^2} + B_1 \frac{\partial^2 U}{\partial v^2} + a_1 \frac{\partial U}{\partial u} + b_1 \frac{\partial U}{\partial v} + c_1 \frac{\partial U}{\partial s} = -4k\Omega^2.$$

The coefficients here are analytic functions of u, v, and s, and the non-vanishing of Ω, which occurs at every regular point, is precisely the condition needed to satisfy the hypotheses of the Cauchy–Kovalevskaya theorem. The result is a very neat application of the theorem, in that the hypotheses needed for Bruns' lemma are precisely those of the Cauchy–Kovalevskaya theorem.

8.3. Commentary

This paper was published posthumously by Mittag-Leffler. Kovalevskaya had apparently communicated it at a meeting in Stockholm, but it is not certain that she intended to try to publish it. Since she did lecture on potential theory at Stockholm in the spring of 1886, it is possible that she discovered this result then but did not have time to communciate it, being busy with her work on the Euler equations. On the other hand the result is exactly the one which Weierstrass says was done while she worked on her dissertation, i.e., an application of the Cauchy–Kovalevskaya theorem to prove that the potential of a suitably shaped body is analytic at the boundary (cf. Section 2.7). Considering the fact that every reference in the paper is to work done before 1874, one is inclined to say that this paper was extracted from her student notes and dusted off for presentation at the meeting. In any case it is certain that her original dissertation contained a result essentially identical to this one.

CHAPTER 9

Evaluations

9.1. Introduction

As the first woman to be admitted to mathematical circles as a regular member, Kovalevskaya naturally elicited more evaluations than any male mathematician would have been subject to. Similarly, although no one would claim that there is any difference between mathematics as practiced by a woman and mathematics as practiced by a man, her works were nevertheless scrutinized much more carefully than those of a male mathematician would have been, simply because, no doubt, people found it fascinating that a woman could actually do such a thing. Opinions on her work at the turn of the century ranged from adulation to disparagement, as will be seen below. For the past half-century the judgment seems to have held pretty much constant, so that one may fairly say that a consensus has been reached. Since that consensus is one to which I myself adhere, it will be given in the final section. The first part of this chapter is devoted to the comments of some distinguished mathematicians on the subject of Kovalevskaya.

9.2. Poincaré

Kovalevskaya's friend Poincaré made frequent and admiring mention of her. His comment on the Cauchy–Kovalevskaya theorem in his paper on the three-body problem (1890) was, "Kovalevskaya significantly simplified the proof and gave it a definitive form." It was pointed out in Section 3.13 that Poincaré wrote a paper on degenerate Abelian integrals about the same time that Kovalevskaya's dissertation on the subject was published. In his analysis of his own works (*Oeuvres* IV, p. 291) he said that it was the work of Kovalevskaya which reawakened his interest in such questions. Similarly, as mentioned in Section 4.10, Poincaré's 1885 paper on hydrodydnamics used techniques very much like those Kovalevskaya had used in her paper on the shape of Saturn's rings. Poincaré himself pointed out the similarity.

9.3. Markov

As we have already seen (Chapter 5), the mathematician A. A. Markov, whose work in probability still ranks very high, dissented from the general acclaim which Kovalevskaya received for her prize-winning paper on the Euler equations. As mentioned in that chapter, Markov doubted her argument that the only cases in which the solutions can be meromorphic are her own and those of Euler and Lagrange. It will be recalled (Chapter 7) that this argument rests on the vanishing of a 6-by-6 determinant for five different values of a parameter. Markov believed that the possibility of multiple roots for the characteristic polynomial should have been considered. At Markov's instigation Lyapunov undertook an investigation of this possibility and showed conclusively (1894) that these three cases were in fact the only ones in which the solutions are single-valued functions of time for all possible initial values. Although one cannot fault Markov for pointing out a gap in Kovalevskaya's reasoning, he gave his discovery more importance than it deserved. After all, the *main* point of Kovalevskaya's memoir was the explicit solution of the equations in a new case. The claim that no other cases exist in which such a solution can be found was relatively unimportant. Markov's target may have been not Kovalevskaya but her friend Chebyshev. He is said to have remarked that his purpose in raising his objections in public was to prove that Chebyshev, who had recommended Kovalevskaya for membership in the Academy of Sciences, had not actually read her memoir (cf. Kovalevskaya's letter to Mittag-Leffler quoted in Section 5.14).

9.4. Kovalevskaya

In an article which appeared in May 1890 (Stillman 1978, pp. 213–228), Kovalevskaya gave her own evaluation of her mathematical career, saying in part:

> [My three dissertations] were adjudged sufficiently satisfactory for the university, contrary to its established procedure, to exempt me from the requirements of an examination and public defense of my dissertation (which is essentially no more than a formality) and to award me directly the degree of Doctor of Philosophy, summa cum laude.
>
> At the same time the first of the works mentioned above was published in Crelle's *Journal für die reine und angewandte Mathematik.* This honor, given to very few mathematicians is particularly great for a novice in the field, inasmuch as Crelle's Journal was then regarded as the most serious mathematics publication in Germany. The best scientific minds of the day contributed to it, and such scholars as Abel and Jacobi had published their work in it in former times. My paper on astronomy, on the form of Saturn's rings, was not published until many years later, in 1885, in the journal *Astronomische Nachrichten.* . . . . . It was at that time [1882] that I embarked on a major new work on the refraction of light in crystals. In the field of mathematics in general it is mostly by reading the works of other scholars that one comes upon ideas for one's independent research. Thus I, too, was led to this topic by studying the work of the French physicist Lamé.
>
> My work was completed in 1883 and had something of an impact in the mathematical world, for the problem of light refraction had not yet been satisfactorily clarified, and I had viewed it from a different, entirely new standpoint.
>
> In addition to the paper on the refraction of light in crystals, several other articles of mine have appeared in *Acta Mathematica,* including (in 1883) the second of my doctoral dissertations, originally presented in 1874 at the University of Göttingen, (on the reduction of a certain class of Abelian functions to elliptic functions.)

All of my scholarly work is written in German or in French. I am as much at home in them as I am in my native Russian . . . . In that same year of 1888 the Paris Academy of Sciences announced a prize competition for the best essay "Sur le problème de la rotation d'un corps solide autour d'un point fixe," [Actually Kovalevskaya gave a Russian translation of the title of the competition topic here, at least in the version printed in Shtraikh (1951). The title given here was the title of her memoir on the subject, which is possibly the way she remembered the competition.] with the proviso that the essay must substantially refine or supplement findings previously attained in this area of mechanics . . . . I sent my manuscript to Paris. By the rules of the competition it was to be submitted anonymously. Therefore I noted down a maxim on the manuscript and attached to it a sealed envelope containing my name, with the same maxim inscribed above it. This was the procedure by which authors retained their anonymity while their work was being evaluated.

The result was beyond what I had hoped. Some fifteen papers were present[ed] but it was mine which was found deserving of the prize. And that was not all: in view of the fact that the same topic had been assigned three times running and had remained unsolved each time, and also in view of the significance of the results achieved, the Academy voted to increase the previously announced award of 3,000 francs to 5,000.

The envelope was then unsealed, and it was learned that I was the author of the work. I was informed immediately and left for Paris to be present at the session of the Academy of Sciences set for the occasion. I was given a highly ceremonial reception and seated next to the president, who made a flattering speech; all in all, honors were lavished upon me.

As I have already mentioned, I have been living in Sweden since 1883 and have become so assimilated to its style of life that I feel completely at home there . . . . I have a large circle of friends and lead an active social life. I am even received at Court . . . .

In reading this passage, one must bear in mind that Kovalevskaya was speaking publicly and therefore diplomatically. As usual in such cases, she was also presenting herself in the best possible light. Thus she stressed her exemption from the usual oral examination without mentioning that she had petitioned for the exemption. Her remark that the rotation problem had been posed "three times running" is evidently an allusion to the Prussian Academy's competitions 1852–1858 mentioned in Section 7.4. The Bordin Prize does not seem to have been offered previously for this problem. Similarly she emphasized the anonymity of the Bordin competition, when in fact it almost certainly was known to the judges which paper was hers. She describes herself as being completely at home in French and German, when actually her German was a source of amusement to both herself and her German friends. She says she is completely at home in Stockholm, when in fact she wanted nothing so much as the chance to move somewhere else. She says that she is received at Court, but does not mention that she made the arrangements to be received on her own initiative, etc.

Nevertheless her praise of her own mathematical papers is quite accurate, with one exception. It *was* a very great honor to have her dissertation published in Crelle's *Journal,* and the work on the rotation problem was deservedly given an increased prize because of its merit. The one exception, of course, is her paper on light refraction, which, as we saw in Chapter 6, is wrong. If this paper had indeed had the impact which she believed it to have, someone would have noticed that in fact it does not give a solution of the equations it claims to solve. Again, it should be noted that although her work was occasioned by reading Lamé, it was unquestionably Weierstrass who posed both the problem and the method to her.

9.5. The Moscow Mathematicians

On 19 February (3 March) 1891, immediately after Kovalevskaya's death, speeches were made in her honor at a meeting of the Moscow Mathematical Society. The mathematician A. I. Stoletov read a biographical essay; N. E. Zhukovsky discussed her applied mathematics; and P. A. Nekrasov discussed her pure mathematics. These speeches, with explanatory notes, were published in the *Math. Sbornik* and later as a separate pamphlet in her honor. Zhukovsky first discussed the paper on the shape of Saturn's rings and assessed its importance quite accurately.

He then turned to her paper on refraction of light and again gave an accurate summary. He expressed the opinion that her solution was superior to Lamé's, but regretted that the solution was not accompanied by any geometrical study to aid in visualizing the solution.

Finally he turned to what he called her principal scholarly glory, the work on the rotation problem. He accompanied this solution with sketches of three different kinds of spinning tops illustrating the cases of Euler, Lagrange, and Kovalevskaya. His conclusion is significant:

> In the summer of 1889 I met Poincaré in Paris, who told me that S. V. Kovalevskaya was working on an extension of the case just discussed and hoped to solve the problem of motion in the case where the center of gravity lies in the equatorial plane of the ellipsoid of inertia for an *arbitrary ellipsoid of rotation* . . . .

If this third-hand report is accurate, Kovalevskaya was trying to eliminate the restrictive hypothesis $A = B = 2C$ from her theorem and solve the problem under just the hypothesis $z_0 = 0$. Such a result would be an enormous advance, and might even have led to a new class of functions, which no doubt would have been called Kovalevskaya functions—the ones which express the solution in the general case.

Nekrasov's essay on Kovalevskaya's contribution to pure mathematics began by citing words which he attributed to the eighteenth-century mathematician Montucla in reference to Montucla's contemporary Maria Gaetana Agnesi, who eventually gave up mathematics and entered a convent. "One cannot but be amazed at a person of the sex so ill-equipped to wrestle with the thorns of science who has penetrated so deeply into all parts of analysis, both ordinary and transcendental." The comparison, so Nekrasov believed, was very suitable, since both women had been taken from the world of learning at an early age. He praised Kovalevskaya for working in both applied and pure mathematics and thereby preserving the tradition of the old school of mathematicians. In his discussion of her work on differential equations, though he warmly praised her work, he chided her slightly for over-emphasizing Weierstrass' contributions to the subject and underestimating those of Cauchy. But, he said, "In the final analysis S. V. Kovalevskaya gave a definitive form to theorems on the integration of partial differential equations, which left nothing to be desired for precision of expression and rigor and simplicity of proof." Very high praise indeed!

After a very thorough discussion of her work on differential equations which proved that he must have read the work in great detail, Nekrasov gave a discussion of her paper on Abelian integrals which showed that he had barely looked at it. He said nothing about the Weierstrass theory of trans-formations on which the paper was based, missed the main theorems entirely, and gave out as the major result an exceptional case of the main theorem which Kovalevskaya had reserved for the end of the paper. Despite this oversight, however, Nekrasov conveyed accurately the significance of the paper. He wrote, "Although these results found by S. V. Kovalevskaya, being special cases, do not have such wide general interest as the results she attained in the theory of differential equations, nevertheless the talent of the author and her ability to penetrate the most complicated relations of analysis are confirmed in a striking manner even here."

Having apparently finished his discussion of Kovalevskaya's work in pure mathematics, Nekrasov nevertheless had much more to say, especially about her work on the rotation problem. Once again his insight is remarkable. He said that the Kovalevskaya case

> is not merely an accidental, fortunate find. On the contrary, this discovery is the result of her persistent stubborn toil and deep knowledge in the field of pure mathematics and especially modern analysis. . . This thought, confirmed by the entire content of her memoir . . . permeates especially the first chapter of it. The introductory chapter in the memoir, which provoked objections from one of the Petersburg mathematicians and which, not playing any essential role in the memoir, is practically excess baggage, nevertheless is very interesting *as the trace of the original path* followed by the mind of S. V. Kovalevskaya in her discovery.

Once again Nekrasov repeated the theme which so impressed other mathematicians of the time about Kovalevskaya's work.

> On the one hand results of this kind show us that even now, with all its complexity, modern analysis can still serve to explain the phenomena of nature. On the other hand such results. . .give us the firm conviction that the significance and importance of modern analysis is inseparable from its complexity, which, no doubt, is only a reflection of the complexity in nature itself.

Finally Nekrasov rendered a comprehensive judgment on Kovalevskaya as a mathematician:

> I do not wish to exaggerate the dimensions of her mental gifts. I say plainly that
> S. V. Kovalevskaya was by no means one of those geniuses of the mathematical
> sciences who by the introduction of fruitful new methods have brought about
> great changes and reforms in various areas of mathematics. But I must say that
> S. V. Kovalevskaya is indisputably the equal of the very talented male mathe-
> maticians of our time, since she penetrated deeply into the available methods
> of science, used them in the most elegant manner, and extended and developed
> them, making completely new, brilliant discoveries and easily dealing with the
> most formidable difficulties.

But Nekrasov could not let well enough alone. After this beautiful tribute to
Kovalevskaya's talent, he upbraided her severely for being a *zapadnitsa*
(Westerner). He said

> It could be plainly said that in her scholarly activity S. V. Kovalevskaya stood
> nearer to Berlin, Stockholm, and Paris than to Moscow or Petersburg. All of
> her learned works are written in foreign languages.

This surely was an unfair, not to say stupid, remark. Kovalevskaya had
made strenuous efforts to obtain an education and a position in Russia, and
she had been rejected at every turn. As for her works being written in foreign
languages, this was the case with most of the great Russian mathematicians.
The journals of the Petersburg Academy were published in French for de-
cades, as the Memoirs of the Prussian Academy in Berlin had once been. The
reason was simple: Scholars wish to have their works read, and French was
the most widely read language of the time. Papers by Russian scholars
continued to be published in French and German, even in Russian journals,
right up to World War II; for instance, Gelfand's *Normierte Ringe* (1941), to
name one known to me.

9.6.   Volterra

Kovalevskaya's reputation, which was very high at the time of her death,
suffered a setback a few months afterwards. On 3 June 1891 Vito Volterra
wrote the following to Mittag-Leffler:

> Monsieur et cher ami,
>
> I cannot begin my letter without expressing to you my deepest regret for the loss
> which the *Acta,* the University of Stockholm, and the mathematical sciences
> have suffered through the death of Mme. Kowalevski. This very cruel loss
> unfortunately precedes only by a few days a very sad communication which I
> consider it my duty to make to you before anyone else, and which I think will
> quite amaze you.
>   In studying for the theory of elasticity, which I wished to expound in my
> lectures, the memoir of Mme. Kowalevski, I verified that the functions which
> she gave as integrals of Lamé's equations on light waves in doubly-refracting
> media (*Acta Mathematica* 6, p. 249) do not satisfy the differential equations.

To convince you of this very strange result, I ask you to examine the formulas given on page 297. It is quite easy to recognize that the functions ξ, η, ζ and the derivatives $\partial\xi/\partial t$, $\partial\eta/\partial t$, $\partial\zeta/\partial t$ vanish when $t = 0$. Since Lamé's equations are linear differential equations, a system of integrals which vanish (together with their derivatives with respect to t) for $t = 0$ would be always zero, which is not the case for the functions ξ, η, ζ.

After that it suffices to assume in these formulas $f(x, y, z) = y$ or $f(x, y, z) = z$, or $f(x, y, z) = z^2$, etc. and to carry out the very simple computations necessary to verify that one finds for ξ, η, ζ functions which do not satisfy the differential equations . . .

Volterra went on to show that Lamé's solution was also wrong and that Kovalevskaya's error arose from forgetting that the Weber coordinate u_2 (cf. Chapter 6) is discontinuous on the wave surface. (Like longitude on a sphere, it must have a saltus along a line analogous to the international dateline unless it is allowed to be a multivalued function.)

Volterra's paper on the subject was printed in the *Acta Mathematica* (1892). As his paper shows, Kovalevskaya was on the right track, but slipped off the rails at the last moment. In fact Volterra gave correct general solutions of the differential equations in a form very similar to Kovalevskaya's:

$$u' = \iint \psi_1(x - \xi, y, z - \zeta)f(t + u_1, \xi, \zeta)\,d\xi d\zeta,$$

$$v' = \iint \psi_2(x - \xi, y, z - \zeta)f(t + u_1, \xi, \zeta)\,d\xi d\zeta,$$

$$w' = \iint \psi_3(x - \xi, y, z - \zeta)f(t + u_1, \xi, \zeta)\,d\xi d\zeta,$$

where again f is arbitrary, u_1 is what Kovalevskaya called λ, and ψ_1, ψ_2, ψ_3 are determined so as to satisfy the equation. The double integrals are extended over a nappe of the wave surface.

9.7. Women in Mathematics

The feminist movement, of which Kovalevskaya was a part and to which her achievements gave strong support, had, by the 1890s managed to open some doors to women, or at least a debate on the desirability of opening doors to women. A young California woman, Dorothea Klumpke, who had gone to France in 1879 to study music, became interested in astronomy. By coincidence she, like Kovalevskaya, studied the rings of Saturn. By another remarkable coincidence, when she received her degree from Sorbonne in 1893, two of her examiners were Kovalevskaya's friend and collegue, Gaston Darboux, and the astronomer Tisserand, who had expressed his admiration for Kovalevskaya by including her paper on Saturn's ring in his treatise. A newspaper clipping saved by Mittag-Leffler reported that Darboux said, "A few years ago the Academy of Sciences, on the report of a committee of which I was a member, gave its highest prize to Mme. Kowalevska. It was a prize that places her name in the same line as the names of Euler and Lagrange in the history of the discoveries relative to the theory of the movement of a solid body aroung a fixed point." (This clipping was dated 15 December 1893. I regret that the publication was not legible on the photocopy which I made. The language of the article was English. Klumpke, I believe, would be an interesting subject for a biography. She apparently received the French Legion of Honor in 1934. I have not seen any obituary of her, but her biography up to age 45 is in the *National Cyclopedia of American Biography* 1906, Volume XIII, p. 377.) A young professor of physics at Clark University, Arthur Gordon Webster, who had studied with Kovalevskaya in Stockholm and later became famous for experimental work on the period of electrical oscillations, wrote to *The Nation* in January 1894 asking when American universities would follow the lead of Europe and admit women on an equal basis with men. Although one might debate the exact moment when women were admitted "on an equal basis" on either side of the Atlantic, the first woman to have matriculated seems to have done so at the University of Vermont in 1871.

In the fall of 1893 the German government allowed women to enter the University of Göttingen, and two years later the first woman to receive a doctorate in mathematics through regular channels was Grace Chisholm Young, who is still remembered for her result on the derivates of an arbitrary function (1916, cited in *Riesz-Nagy* 1955, p. 17). Mrs. Young's doctoral supervisor was Felix Klein. A moving account of Mrs. Young and her husband W. H. Young was written by Grattan-Guinness (1972), who points out that Klein's supervision came about as a result of an initiative to further women's higher education taken by the Prussian Kultusminister, Friedrich Althoff. Althoff's biography in Poggendorff's *Biographische-Litterarisches Handwörterbuch* mentions Klein as one of Althoff's allies, and says that one of Althoff's last acts was to unify men's and women's education.

9.8. Loria

The achievements of the women just discussed were discounted by certain champions of male supremacy. One of the latter was the mathematician Gino Loria, whose reputation was sufficiently high that well-chosen words from him might do real harm. In 1904 he wrote an article for the *Revue Scientifique* in which he expressed the hope that the mathematical faculties would remain closed to women. His response to the objection that women like Sophie Germain and Kovalevskaya had proven themselves capable of producing good mathematics was the following:

> As for Sophie Germain and Sonja Kowalevsky, the collaboration they obtained from first-rate mathematicians [Sophie Germain corresponded with Gauss] prevents us from fixing with precision their mathematical role. Nevertheless

what we know allows us to put the finishing touches on a character portrait of any woman mathematician: She is always a child prodigy, who, because of her unusual aptitudes, is admired, encouraged, and strongly aided by her friends and teachers; in childhood she manages to surpass her male fellow-students; in her youth she succeeds only in equalling them; while at the end of her studies, when her comrades of the other sex are progressing, fresh and courageous, she always seeks the support of a teacher, friend or relative; and after a few years, exhausted by efforts beyond her strength, she finally abandons a work which is bringing her no joy . . . .

One can only wonder what biographies of Kovalevskaya Loria read, that he believed she was "admired, encouraged, and strongly aided by her friends and teachers" as a child, or that she had finally abandoned "a work which is bringing her no joy."

The most pernicious part of Loria's article, however, is not his travesty of the education of mathematically gifted girls. It is the implied major premise that one must be able to "fix with precision" the amount of originality in the works of a woman mathematician before one can admit she is good. Would he like this criterion applied to his own works? In fact, if a person has the right to put his or her name on an article, then that person should receive credit for the work. Everyone uses the results of others, but Kovalevskaya was scrupulous about citing Weierstrass where she used his ideas. (In fact, one might believe she was overscrupulous, and thus did not receive credit for any clarifications she might have added to Weierstrass' work.) Loria's view is plainly the result of blind prejudice. Otherwise he could not have failed to see how ludicrous it is to compare the career of a male mathematician, who from childhood was expected to have a profession and was given an education with this aim in mind, with that of a female mathematician, who received no encouragement from her parents and had to go to extraordinary lengths to obtain even a partial and highly informal education without any hope of a career.

Fortunately nothing approaching Loria's crude antifeminism has been heard lately. The *fact* that women can do mathematics was firmly established by Kovalevskaya and amply confirmed by the many women who have enriched twentieth century mathematics. This fact remains, irrespective of what psychologists may later find out about the differences, if any, in the way men and women think.

9.9. Mittag-Leffler

In 1923 Mittag-Leffler devoted an entire volume of the *Acta Mathematica* to Kovalevskaya, Weierstrass, and Poincaré. In this issue he published many of Weierstrass' letters to Fuchs, du Bois-Reymond, and Kovalevskaya, together with two long articles, one on Weierstrass' first 40 years, and one on Weierstrass and Kovalevskaya. These articles are good primary sources for those interested in the persons named and have been extensively used in the preparation of this book. The material is largely factual, however. For Mittag-Leffler's own judgment on Kovalevskaya one must consult his obituary of her (1892).

> Sophie Kovalevsky will retain an eminent place in the history of mathematics, and her posthumous work [apparently Nigilistka], which should soon appear, will keep her name in the history of literature. But it is perhaps neither as mathematician nor writer that one should properly appreciate or above all judge this woman of so much spirit and originality. As a person she was even more remarkable than one would be able to believe from her works. All those who knew her and were near to her, to whatever circle or part of the world they belonged, will remain forever under the lively and powerful impression which her personality produced.

9.10. Klein

In his posthumous treatise (1926) on nineteenth century mathematics, Felix Klein devoted two pages to Kovalevskaya as part of his discussion of Weierstrass. An unfortunately large share of his comments were negative, though not in themselves unfair. He mentioned, for instance, Volterra's discovery of the error in her paper on the Lamé equations, and said that one was "not completely satisfied" with the work on the rotation problem. Evidently he was alluding to the controversy stirred up by Markov.

After these comments, however, Klein expressed great admiration that Kovalevskaya achieved so much in such a short and turbulent life. In a comment which feminists will no doubt find particularly irritating, he said that the mathematical community ought to thank Kovalevskaya for coaxing Weierstrass out of his reserve.

These comments, read casually, leave the impression that Klein was denigrating Kovalevskaya's value, particularly as he said that, since her works were all written in Weierstrass' style, one could not be sure how much of the work was her own. This impression is entirely misleading, since Klein did express admiration for her. His negative comments are the result of serious objections raised by competent mathematicians. He could not, in the space of the two pages he felt justified in devoting to Kovalevskaya, give a detailed analysis of all of her work, showing what seem to be her own ideas as opposed to those of Weierstrass and examining the implications of the issues raised by Volterra and Markov.

Particularly among mathematicians Klein's book has been enormously influential. It is unfortunate that its effect on Kovalevskaya's reputation has been harmful, and doubly ironic that this harm should come from Klein, who did so much to advance the position of women. Klein was certainly no champion of male supremacy. In an article in *Le Progrès de l'Est* (Nancy 28 March 1898) he is quoted as saying that he found his women students to be in every respect the equals of their male colleagues.

9.11. Golubev

Soviet scholars have understandably studied Kovalevskaya more thoroughly
than Western mathematicians and have generally found much to praise.
Professor V. V. Golubev used the Kovalevskaya case of a rotating rigid body
as the theme for a course in mathematical physics at Moscow State University
in 1950. His lectures were subsequently translated into English (Golubev
1953). Golubev also wrote an article on the subject, giving a more concise
statement of what he considered to be the merits of the work (1950):

> . . . attempts were made by her to solve the problem by inverting hyperelliptic
> integrals, and along with that, the whole approach to solving the problem took
> on that original character, unexpected in its boldness of thought, which always
> impresses the reader of her works . . .
>   Despite a series of works of N. E. Zhukovsky, G. G. Appel'rot and other
> researchers, the general picture of the motion remains so mysterious that pro-
> posals have been heard to resort to so heroic a measure as the experimental
> study of the motion of a top corresponding to the hypotheses of the Kovalev-
> skaya case. All that we know at present reduces to the fact that the motion in
> the Kovalevskaya case is immeasurably more complicated than that which we
> have in the cases of Euler–Poinsot and Lagrange–Poisson.
>   It is known that shortly before her premature death S. V. Kovalevskaya
> indicated in a conversation with Poincaré that she had found numerous other
> cases where the problem can be solved completely. Combining the ideas of
> S. V. Kovalevskaya which led her to the solution of the problem in the case she
> indicated with this assertion, one might conjecture that the method which she
> used is connected with the theory of algebraic functions and integrals.

9.12. Kovalevskaya as a Mathematician

Since Chapters 2–4 and 6–8 of this book were written to enable the reader to
judge for himself or herself what kind of mathematician Kovalevskaya was,
this final section will give only a brief statement of my own estimate of her
work. It seems to me that the total corpus of her work presents us with a
portrait of a competent, creative mathematician who produced some valuable
work and a few works of minor importance, making an occasional mistake
in the process. In short she was a mathematician like most others, more gifted
than most, but distracted from her work far more often as well. As Nekrasov
correctly points out, she was not one of the giants of nineteenth-century
mathematics, but then neither were most of her male colleagues, who nevertheless made substantial contributions to the subject. Since she was a student
of Weierstrass, there are plenty of mathematicians with whom she could be
compared, if comparisons are deemed desirable. Though she was not
Weierstrass' best student, she certainly could hold her head up in such
company. But in any case, such comparisons are unfair, since Weierstrass'
other students did not have the handicaps to overcome that she had. For the
same reason she has justifiably been given more attention than other
mathematicians—Jacobi, for instance—whose contributions were much

greater. It is certain that she would not seem quite so remarkable if she lived a century later, when no one is surprised to see a competent woman mathematician. What one must remember is that this easy acceptance of what women can do is in no small part due to Kovalevskaya's achievements; and at least some of those achievements are grand on an absolute, not relative, scale. I would like to close by correcting what seem to me to be misimpressions left by many previous accounts of Kovalevskaya's works.

First of all, a contrast is sometimes made between her supposedly "applied" interests and Weierstrass' "pure" mathematics. It is asserted that this contrast proves her independence of Weierstrass. I believe this view is based on a misunderstanding of what her papers actually say. For example, the fact that her prize-winning memoirs mentioned the problem of a rigid body rotating about a fixed point seems to imply that these papers were in some sense mathematical physics. Actually the problems were entirely mathematized long before Kovalevskaya worked on them. Her interest was the purely mathematical one of finding the solution of a set of differential equations by means of theta functions. Exactly the same comments apply to her work on "light refraction." This problem was a problem in physics at the time of Fresnel; Lamé had made it entirely mathematical before it got to Kovalevskaya; and her interest had been in solving a system of differential equations using a technique of Weierstrass. The titles of Chapters 6 and 7 in this book were deliberately chosen to convey the real essence of her work.

In this connection it should be noted that all of her "applied" papers were written at Weierstrass' suggestion, except possibly the paper on Saturn's rings. In fact it appears that Kovalevskaya's interest in applications was almost identical to that of Weierstrass. They formed an interesting frame or background for the mathematics which occupied the foreground. Weierstrass frequently lectured on the applications of mathematics and encouraged his students to study problems connected with physics. With the possible exception of his paper on dioptrics (*Werke* III, pp. 175–178) his interest in applications seems to have been as "a rich collection of examples for his methods," to quote an attitude he criticized in his *Antrittsrede* at Berlin.

Definitions of what constitutes applied mathematics vary. As a basis for discussion, the following list of seven modes of research in mathematics and physics (Grattan–Guinness 1981, p. 110) may help to clarify the role of Kovalevskaya's work:

1. Mathematics pursued with no apparent interpretations in physics in mind (and none sought), although it may play a role in interpretable mathematics in some way.

2. Mathematics pursued with no interpretation in physics involved, but with an awareness that such interpretations exist. Indeed, they may have motivated the mathematics in the first place.

3. Mathematics pursued with an interpretation in physics initially in mind, but where the guidance of the physical problem has been lost. The result is, for

example, a mathematical expression for a physical effect which cannot possibly be currently detected, or be computed from that expression with an error less than the effect itself; or the use of mathematical relationships between physical constants which make no physical sense; or the use of a hopelessly oversimplified physical model in the first place; or methods or reasoning in which physical concepts play little or no role.

4.   Mathematics pursued with an interpretation in physics in mind and where the guidance of the physical problem provides adequate control on the mathematics produced. The rigor of the mathematics may be suspect.

5.   Theoretical analysis of (certain) physical phenomena, where the mathematical component is small or even nonexistent. Such work is often foundational in character, or concerned with physical constants.

6.   Experimental work in physics, designed to test the (mathematical or nonmathematical) theory—including the theory of any instruments involved—that has been worked out already.

7.   Engineering constructs, machines, and instruments: that is the design of equipment and structures to fit particular situations or types of situations.

Probably Kovalevskaya's paper on light refraction and the rotation problem, despite the fact that H. A. Schwarz and N. E. Zhukovsky gave descriptions of physical models of the Kovalevskaya case of a rotating body, are best described by the second and third of these categories. The papers were definitely not written with a view to helping physicists understand any physical phenomena. Indeed the elastic solid theory of light was rapidly being replaced by the electromagnetic theory at the time when Kovalevskaya wrote her paper on light refraction.

Kovalevskaya's paper on Saturn's rings definitely fits the fourth model. The rigor of the mathematics is not exactly in question, but that is because no theorems are stated. Instead Kovalevskaya suggests techniques which someone *might* be able to formulate into a rigorous theorem under some additional hypotheses. Her computations are not feasible, except in very special cases.

In general it should be pointed out that the problems Kovalevskaya worked on had already been formulated as problems in pure mathematics before she got them (with the exception of the work on Saturn's rings) and that she made no attempt to show how her results apply to the physical problem which originally gave rise to the problems she worked on. For these reasons I do not consider it accurate to describe her as an applied mathematician.

Now I would like to summarize the specific points on which my judgment of each of her works differs from previous judgments.

(1)   Previous expositions of Kovalevskaya's theorem in partial differential equations have formulated the theorem in modern language and notation, thereby making it appear to be an existence and uniqueness theorem. In fact, however, the modern notions of existence and uniqueness were still being developed in 1874. Kovalevskaya's own language makes it clear that she thought she was proving that a differential equation could be used as the definition of an analytic function. This connection of the theorem with its

birthplace in the Weierstrass theory of analytic functions should be kept in mind.

(2) To understand Kovalevskaya's mathematics it is necessary to know the topic which was her particular area of expertise, Abelian integrals. Yet no one has ever quoted even one theorem from her paper on this subject. (Even the apparent exception—Nekrasov, as discussed in Section 9.5—was actually misquoting her). Kovalevskaya's approach to the prize-winning topic of a rotating rigid body cannot be understood without a thorough understanding of the influence this topic had on all her thinking.

(3) Many authors have implied that the major result of Kovalevskaya's paper on Saturn's rings is that the ring has an oval cross section in second approximation. In fact that result is a trivial consequence of her decision to use Fourier series, and she hardly mentions it. Of more importance to her was a quantitative criterion distinguishing whether the oval cross section points toward or away from Saturn. I repeat that the shape of successive approximations depends on the method of approximation and has nothing to do with the physical model. On the other hand, in this paper she raised two important issues in applied mathematics involving error analysis and stability and also proposed, in a heuristic way, certain techniques for solving integral equations, techniques which were rediscovered and rigorously developed by Hammerstein in 1930.

(4) It has often been said that Kovalevskaya wrote an angry letter to Weierstrass in 1884 reproaching him for not finding the error which Volterra pointed out in her paper on light propagation. As we have seen above (Section 9.6) Volterra did not even discover the error until after Kovalevskaya's death. Hence no such letter could have existed. The many writers who make this error, I believe, were misled either by Mittag-Leffler's printing of a letter from Weierstrass to explain why the error was not caught by Weierstrass (*Acta Mathematica* 39, p. 193), or by secondary articles based on Mittag-Leffler's article.

(5) I have stressed above that the real significance of Kovalevskaya's prize-winning work on the rotation of a rigid body was that it applied some very esoteric mathematics to a problem of physics. I would like to add that it also serves as a good example of the analytic, Berlin school of mathematics in contrast to the geometric, Göttingen school of work exemplified by Klein's work on the same topic, which was discussed in Section 7.9.

(6) Kovalevskaya's posthumous article on Bruns' theorem (1891) was very likely a previously unpublished part of her doctoral dissertation, or at least of a first draft of the dissertation.

As this book has been confined to the mathematics of Kovalevskaya, little has been said about her very important literary and social activity. For these aspects of her life the reader is referred to the excellent works by Koblitz (1983) and Kochina (1981). It is a tribute to her talent that even with this large part of her activity abstracted she remains an impressive figure, deservedly, as G. J. Tee (1980) has said, a heroine of the feminist movement. Her achievements established the fact that women can hold their own in mathematics. Among the students of Weierstrass she was mathematically "quite respectable." Her competence and poise as a member of the faculty in Stockholm established that society can, without falling apart, allow women to become the colleagues of men on university faculties. Although the realization of all her ambitions for women in mathematics is not yet complete, it continues apace. The impetus she gave to this movement a century ago and the permanent advances in knowledge which her works gave to the world form the enduring foundation of her fame.

PART III

APPENDICES AND BIBLIOGRAPHY

Appendices

These appendices serve two different purposes. Appendices 1, 2, 4, and 5 are written to help the undergraduate reader who has not yet reached some of the mathematics used in the main part of the text. The reader is expected to know roughly two years of undergraduate mathematics. Since Kovalevskaya's mathematics goes somewhat beyond that level, particularly into complex analysis, and since some of the other work described in this text, particularly that of Euler and Lagrange, requires a bit of classical mechanics and calculus of variations, these appendices are included for reference. Needless to say, they are not textbooks. Their only purpose is to make things a little less unfamiliar. The other two appendices (3 and 6) describe some topics in differential equations which *apparently* are no longer in the standard curriculum. They are included to satisfy the curiosity of mathematicians.

Appendix 1.   The Method of Majorants

A useful technique for solving an initial-value problem

$$u'(x) = f(u(x), x)$$
$$u(0) = u_0$$

is to assume a power series expansion $u(x) = \sum_{n=0}^{\infty} u_n x^n$ and use the equation itself to calculate $u_1, u_2, \ldots$. For example, the problem

$$u'(x) = u(x)$$
$$u(0) = 1$$

leads to the equations

$$u_0 = 1$$
$$n u_n = u_{n-1}.$$

It is easy to prove by induction that $u_n = 1/n!$, so that $u(x) = \sum_{n=0}^{\infty} (1/n!)x^n = e^x$, which proves that $u(x) = e^x$ is the only analytic solution of this problem.

In general this method of undetermined coefficients tells only what the Taylor coefficients of a solution must be if there is an analytic solution. To verify that there actually is an analytic solution, one must prove that the series obtained from the method converges. The following sequence of lemmas and theorems is intended to do that for the general initial-value problem posed above.

Lemma. *For all integers i, j, k with $k \geq 0$, define polynomials $P_{i,j,k}(z_0, \ldots, z_k)$ in $k + 1$ variables $z_0, \ldots, z_k$ as follows:*

$$P_{0,0,0}(z_0) = 1$$

$$P_{i,j,0}(z_0) = 0 \qquad \text{for all other values of } i \text{ and } j$$

$$P_{i,j,k+1}(z_0, \ldots, z_{k+1}) = P_{i,j-1,k}(z_0, \ldots, z_k) + z_1 P_{i-1,j,k}(z_0, \ldots, z_k)$$

$$+ \sum_{r=0}^{k} z_{r+1} D_r P_{i,j,k}(z_0, \ldots, z_k).$$

Then the polynomials $P_{i,j,k}$ have the following properties:

(1) *The coefficients of $P_{i,j,k}$ are non-negative integers;*
(2) *if $u(x)$ and $f(u,x)$ are continuously differentiable up to order n, the composite function $g(x) = f(u(x), x)$ satisfies*

$$g^{(n)}(x) = \sum_{j=-\infty}^{\infty} \sum_{i=-\infty}^{\infty} P_{i,j,n}(u(x), u'(x), \ldots, u^{(n)}(x)) D_1^i D_2^j f(u(x), x).$$

The terms for which $i < 0$ or $j < 0$ or $i + j > n$ all vanish.

Proof. Assertion (1) has a simple inductive proof, as does the fact that the terms are zero when the indices are in the indicated ranges. It remains only to show that $g^{(n)}(x)$ is equal to the indicated expression. This assertion is obvious for $n = 0$. Supposing it true for $n = k$, we differentiate both sides with respect to x. The inductive definition of $P_{i,j,k+1}$ is then precisely the formula which shows that the relation holds for $n = k + 1$. No convergence difficulties occur since the supposedly infinite sum is in fact finite, due to the vanishing of most of the terms.

Corollary. *For all integers i, j, n with $n \geq 0$, there exist polynomials $Q_{i,j,n}(z_0, \ldots, z_n)$ whose coefficients are non-negative such that if $u(x)$ is analytic at $x = 0$ and $f(u,x)$ is analytic at $(0,0)$ and $u(0) = 0$, then the composite analytic function $g(x) = f(u(x), x)$ has a Taylor series $g(x) = \sum_{n=0}^{\infty} g_n x^n$ whose coefficients are given by*

$$g_n = \sum_{i=0}^{n} \sum_{j=0}^{n-i} a_{ij} Q_{i,j,n}(u_0, \ldots, u_n) \tag{1}$$

where $f(u,x) = \sum_{j=0}^{\infty} \sum_{i=0}^{\infty} a_{ij} u^i x^j$ and $u(x) = \sum_{n=0}^{\infty} u_n x^n$.

Proof. Since $a_{ij} = (1/i!\,j!)\, D_1^i D_2^j f(0,0)$ and $u_n = (1/n!)\, u^{(n)}(0)$, while $g_n = (1/n!)\, g^{(n)}(0)$, simply take $x = 0$ in the lemma. It follows that one can (and must) take

$$Q_{i,j,n}(z_0, \ldots, z_n) = \frac{i!\,j!}{n!} P_{i,j,n}(z_0, z_1, 2z_2, \ldots, n!\,z_n).$$

Theorem. *If $u(x)$ and $f(u, x)$ are analytic at 0 and $(0, 0)$, respectively, and have the Taylor coefficients shown above for expansion about these points, and if g_n is defined by (1), then the series $\sum_{n=0}^{\infty} g_n x^n$ converges to the function $g(x) = f(u(x), x)$.*

Proof. It is well known that the composite function $f(u(x), x)$ is analytic; hence its Taylor series converges. By the corollary the coefficients of this series are given by (1).

We are interested in the equation that results when we take $u'(x) = g(x)$, i.e., $u(x)$ satisfies the initial-value problem posed at the beginning of the Appendix. If there is an analytic function $u(x)$ satisfying this problem, its coefficients are determined by the equations

$$u_0 = 0$$

$$(n + 1)u_{n+1} = \sum_{i=0}^{n} \sum_{j=0}^{n-i} Q_{i,j,n}(u_0, \ldots, u_n) a_{ij} \tag{2}$$

Our task is to show that if u_n are defined by this system of equations, then the series does converge. Here the non-negative coefficients of $Q_{i,j,n}$ play an essential role.

We consider an auxiliary problem

$$v'(x) = h(v(x), x)$$

$$v(0) = 0$$

in which $h(v, x) = \sum_{i=0}^{\infty} \sum_{j=0}^{\infty} b_{ij} v^i x^j$ and $|a_{ij}| \leq b_{ij}$ for all i and j. We shall show that $h(v, x)$ can be chosen so that an analytic solution $v(x) = \sum_{n=0}^{\infty} v_n x^n$ exists. From the recurrence relation

$$v_{n+1} = \frac{1}{n+1} \sum_{i=0}^{n} \sum_{j=0}^{n-i} Q_{i,j,n}(v_0, \ldots, v_n) b_{ij}.$$

it is easy to prove that $|u_n| \leq v_n$ for all n. For suppose that

$$Q_{i,j,n}(z_0, \ldots, z_n) = \sum C_{r_0, \ldots, r_n}^{i,j,n} z_0^{r_0} \cdots z_n^{r_n}.$$

We have already observed that the coefficients $C_{r_0, \ldots, r_n}^{i,j,n}$ are non-negative. Therefore if $|u_k| \leq v_k$ for $k \leq n$ (which is certainly the case for $n = 0$), then

$$|Q_{i,j,n}(u_0, \ldots, u_n)| \leq \sum C_{r_0, \ldots, r_n}^{i,j,n} |u_0|^{r_0} \cdots |u_n|^{r_n}$$

$$\leq C_{r_0, \ldots, r_n}^{i,j,n} v_0^{r_0} \cdots v_n^{r_n}$$

$$= Q_{i,j,n}(v_0, \ldots, v_n).$$

A similar argument using the recursion relation for u_{n+1} and v_{n+1} now proves that $|u_{n+1}| \leq v_{n+1}$.

Thus the fact that the series for $v(x)$ converges implies that the series for $u(x)$ converges also. The problem has thus been reduced to finding a suitable majorant function $h(v, x)$ for which the initial-value problem has an analytic solution. To this end, we note that for K, ρ, and σ all positive the choice $h(v, x) = K(1 - v/\rho)^{-1}(1 - x/\sigma)^{-1}$ gives an initial-value problem whose solution is the analytic function

$$v(x) = \rho(1 - \sqrt{1 + 2\rho^{-1}\sigma K \log(1 - \sigma^{-1}x)}).$$

On the other hand, if $f(u, x)$ is analytic, its Taylor series $\Sigma\, a_{ij}u^ix^j$ converges absolutely for some values $u = U$ and $x = X$, both of which are positive, so that $|a_{ij}U^iX^j|$ is bounded for all i and j. Let K be an upper bound for these numbers, and take $\rho = U$ and $\sigma = X$. Then, by the mere definition of K we have $|a_{ij}| \le K\rho^{-i}\sigma^{-j}$ for all i and j. But $K\rho^{-i}\sigma^{-j}$ is precisely b_{ij}, the Taylor coefficient of $h(v, x)$. Thus we have found a suitable majorant and arrived at the following theorem:

If $f(u, x)$ is analytic at $(0,0)$, there exists one and only one function $u(x)$ analytic at $x = 0$ and satisfying the equations

$$u'(x) = f(u(x), x); \qquad u(0) = 0.$$

It is trivial to replace the second of these equations by $u(x_o) = u_o$ when $f(u, x)$ is analytic at (u_o, x_o).

Appendix 2.   Some Complex Analysis

In most courses on advanced calculus students are introduced to the classical Riemann integral $\int_a^b f(x)dx$, which is said to be approximated by a Riemann sum obtained through the following procedure:

(1)   Choose any $n + 1$ points $x_o, \ldots, x_n$ satisfying

$$a = x_o < x_1 < x_2 < \cdots < x_n = b.$$

(2)   Choose any n points $x_1^*, x_2^*, \ldots, x_n^*$ such that $x_{j-1} \le x_j^* \le x_j$ for $j = 1, \ldots, n$.

(3)   Form the sum $S = \Sigma_{j=1}^n f(x_j^*)(x_j - x_{j-1})$.

Obviously there are many Riemann sums approximating with greater or lesser accuracy a given Riemann integral. When $f(x)$ is continuous on $[a, b]$, the approximation can be made as accurate as desired, independently of the choice of the x_j^*, by taking the points $x_0, \ldots, x_n$ close together, i.e., by taking a suitably fine partition of the interval $[a, b]$. At least in my experience, most students object to this complicated definition of the integral and prefer simply to remember formulas like

$$\int \frac{1}{x}\, dx = \log|x| + C.$$

For the extension to complex numbers, however, it is essential to have an analog of the Riemann definition, since the formula just given is definitely wrong if x is allowed to be a complex number. To carry out this extension is our first task.

Notice that the formula for a Riemann sum would make sense even if the points involved were complex numbers. A complication would arise from the fact that in the plane there are many ways to go from a to b. For that reason we must specify a "path of integration," which we shall label γ and write $\int_\gamma f(z)dz$ instead of $\int_a^b f(z)dz$. Our definition of the approximating sum is like the definition of Riemann sum:

(1)   Choose $n + 1$ points $z_0, \ldots, z_n$ such that z_{j-1} precedes z_j on γ, $j = 1, \ldots, n$.
(2)   Choose any n points $z_1^*, \ldots, z_n^*$ on the path such that z_j^* is between z_{j-1} and z_j.
(3)   Form the sum $S = \sum_{j=1}^n f(z_j^*)(z_j - z_{j-1})$.

As in the case of Riemann integrals, one can show that if $f(z)$ is continuous, there is a unique complex number such that all the approximating sums for sufficiently fine partitions are very close to this number. The number is of course called the integral of f over the path γ shown in Figure A2-1.

In practice, to evaluate the complex integral, one parametrizes the path γ as the image of a line segment: $z(t) = x(t) + iy(t)$ for $0 \leq t \leq 1$ traces out γ as t varies. Then if $f(z) = u(z) + iv(z)$ where u and v are real, the complex integral $\int_\gamma f(z)\, dz$ becomes

$$\int_0^1 u(z(t))x'(t) - v(z(t))y'(t)\ dt + i\int_0^1 u(z(t))y'(t) + v(z(t))x'(t)\ dt$$

Even this simpler procedure is not often practical, and one often evaluates the integral using special theorems developed in complex analysis, the residue theorem, for instance. For theoretical purposes it is much more important to know certain formal properties of the integral than to be able to evaluate it. Some of these properties are the following:

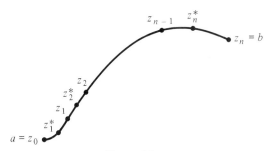

Figure A2-1.

(1) $\int_\gamma f(z)dz$ is a complex number determined by the function $f(z)$ and the path γ.

(2)   If β is a path which begins at the point where γ ends, we can denote the composite path obtained by traversing first γ and then β by $\gamma + \beta$. (Note that in general $\beta + \gamma$ does not make sense.) Then

$$\int_{\gamma+\beta} f(z)dz = \int_\gamma f(z)dz + \int_\beta f(z)dz.$$

(3)   If $-\gamma$ denotes the path γ traced in reverse (from b to a), then

$$\int_{-\gamma} f(z)dz = -\int_\gamma f(z)dz.$$

(4)   The endpoints of the path need not be distinct. In fact an important role is played by closed paths (loops) which begin and end at the same point. Notice that if γ and β are two paths which both begin at a and end at b, then the path $\gamma + (-\beta)$, which we denote by $\gamma - \beta$, is a loop starting and ending at a, as in Figure A2-2, and

$$\int_{\gamma-\beta} f(z)dz = \int_\gamma f(z)dz - \int_\beta f(z)dz.$$

Statement 1 is obvious; statements 2 and 3 are easily proved by considering the approximating sums; statement 4 is a consequence of 2 and 3. If $f(z)$ has the convenient property that $\int_\gamma f(z)dz = 0$ whenever γ is a loop, then $\int_\gamma f(z)dz = \int_\beta f(z)dz$ for any two paths from a to b; for $\gamma - \beta$ is a loop, and the equation of statement 4 comes into play. In this happy case we can without ambiguity use the symbol $\int_a^b f(z)dz$ and evaluate the integral along any convenient path. It would therefore be convenient to have theorems which guarantee that integrals around loops are zero. (They are not always zero, as will be shown below.) One such theorem is the Cauchy integral theorem, which asserts that if $f'(z)$ exists at every point of a region without any holes in it, then such is indeed the case. This theorem applies in particular to $f(z) = z^n$ for non-negative integers n. Then $f'(z) = nz^{n-1}$ at every point, and in fact we even have the familiar formula

$$\int_a^b z^n dz = \frac{b^{n+1} - a^{n+1}}{n + 1}.$$

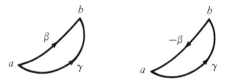

Figure A2-2.

The Cauchy integral theorem does not apply if n is negative, since the region where the derivative exists has a hole in it (the point 0 is missing). Nevertheless, by accident the preceding formula continues to hold for every n with the single exception of $n = -1$, for which the formula does not make sense. All of our labor can therefore be concentrated on the case $n = -1$.

The fundamental fact we have to deal with is that if γ is a path which winds counterclockwise once around 0, then

$$\int_\gamma \frac{1}{z}\,dz = 2\pi i.$$

This relation is easily verified in the special case of a circle by using the parametrization $z(t) = \cos 2\pi t + i \sin 2\pi t$. It holds in general, however. If the path winds around 0 n times the integral is $2\pi n i$. Thus for the path in Figure A2-3 the integral $\int_\gamma (1/z)\,dz = 4\pi i$.

In analogy with the Riemann integral we would like to *define* the logarithm for complex numbers by the equation

$$\log w = \int_1^w \frac{1}{z}\,dz.$$

The difficulty with doing so is that the integrand is not one which gives integrals independent of the path. Hence it appears that we are defining only certain possible values of the logarithm, different values being given by different paths from 1 to w. From what has been said above, it follows that if $\log w$ is a possible value of the logarithm, so is $\log w + 2\pi i$ and even $\log w + 2\pi n i$ for any integer n. As it happens these are the only possible values for the logarithm, and all are equally acceptable, in the sense that for any such value we shall have the characteristic property of the logarithm, i.e., the equation

$$e^{\log w} = w.$$

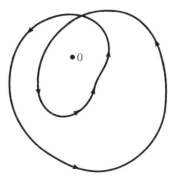

Figure A2-3.

The fact that there are so many possible values of log w is reflected in the fact that $e^{2\pi i} = 1$, i.e., the *periodicity* of the exponential function is reflected in the *multivaluedness* of the inverse function. This principle will recur constantly in what we do. I remark in passing that no sensible value can be given to the expression log 0. For any other value of w we can define log w *locally* so as to be single-valued and continuous, by taking a suitable branch of the log, i.e., a suitable path from 1 to z and requiring that the path from 1 to nearby points not depart very much from the path just chosen. However, if w traces a path around the origin, log w cannot vary continuously and return to its initial value when w does so. We could, of course, cut out the ambiguity by specifying that the path from 1 to w must be a straight line. If we did so, we would find log $w = \log|w| + i\theta$, referring to Figure A2-4.

This definition has the disadvantage that it leaves the logarithm undefined on the negative half of the real line, since the path cannot pass through 0, where $1/z$ is undefined. A further disadvantage is that the formula log $zw = \log z + \log w$ holds for some pairs z and w but not for others.

A more sophisticated way of handling the problem is to imagine that a circuit of 0 moves z from one level to another, like a winding staircase. If we have a stack of sheets representing copies of the complex plane, we can say that crossing the negative real axis from top to bottom moves us up one sheet and crossing it from bottom to top moves us down one sheet. The visual representation is as if we cut each copy of the plane along the negative real axis as in Figure A2-5 and glued the top edge of each cut to the bottom edge of the cut on the sheet above it. We then fix one sheet as a base, which we call sheet 0, and label the other sheets as in the side view of the stack, as shown in Figure A2-6.

This spiral stack of sheets is called the Riemann surface of the logarithm. It extends infinitely high and infinitely low, so that for each complex number except 0 we have infinitely many copies, all regarded as numerically the same but geometrically different: $\ldots w_{-n}, \ldots, w_{-1}, w_0, w_1, \ldots,$ $w_n, \ldots$. We can use our original definition of log w on the Riemann surface

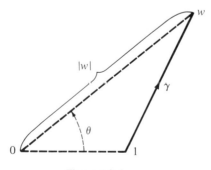

Figure A2-4.

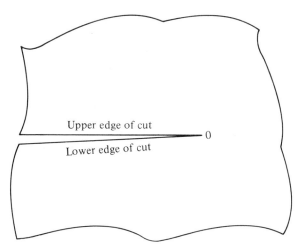

Figure A2-5   Top view when the cut is made.

without ambiguity, since paths cannot wind around zero and return to their starting point. In fact the integral around any closed loop on the Riemann surface $(\int_\gamma (1/z)dz)$ is zero. The formula $\log w_n = \log w_0 + 2\pi ni$ shows how $\log w$ varies among the points on the surface corresponding to a single complex number.

It is important to note that all these complications with the Riemann surface were occasioned by the fact that the formula

$$t = \int_1^w \frac{1}{z} dz$$

does not define t as a single-valued function of w; on the other hand, the equation *does* define w as a single-valued function of t, namely, $w = e^t$. The multivaluedness of t as a function of w is reflected in the periodicity of w as a function of t. This property turns out to be very common in the study of the integrals of algebraic functions. The integrals themselves are badly multi-valued, but the inverse functions are well behaved. We shall investigate this phenomenon by considering the problem of integrating some easy algebraic functions.

Figure A2-6   Side view showing how to glue the sheets together.

Next to the polynomials, whose integrals can be worked out from the formula for integrating z^n, the simplest functions are the rational functions, which consist of a polynomial plus a quotient

$$\frac{c_0 + c_1 z + \cdots + c_m z^m}{d_0 + d_1 z + \cdots + d_n z^n},$$

with $m < n$. Provided the denominator can be factored (which is a problem belonging properly to algebra and not to complex analysis), a rational function also can be integrated, due to the partial fractions decomposition. If the denominator in the fraction just shown factors as

$$d_0 + d_1 z + \cdots + d_n z^n = d_n (z - r_1)^{k_1} \cdots (z - r_j)^{k_j},$$

where $r_1, \ldots, r_j$ are distinct complex numbers and $k_1, \ldots, k_j$ are positive integers, then there exist constants A_{pq}, $1 \le p \le k_q$, $1 \le q \le j$, such that

$$\frac{c_0 + c_1 z + \cdots + c_m d^m}{d_0 + d_1 z + \cdots + d_n z^n} = \sum_{q=1}^{j} \sum_{p=1}^{k_q} \frac{A_{pq}}{(z - r_q)^p}.$$

It follows that the rational function can be integrated and that the integral is a combination of terms of the form $(z - r)^p$ and $\log(z - r)$. These functions are called the elementary functions, and we have therefore just shown that every rational function can be integrated in terms of elementary functions.

We can do still better, however. If the integrand involves the nth root of a linear function, i.e., $\sqrt[n]{az + b}$, the change of variable $w^n = az + b$ [i.e., $z = a^{-1}(w^n - b)$] converts $\int F(z, \sqrt[n]{az + b})\, dz$ into

$$\int F\left(\frac{w^n - b}{a}, w\right) \frac{nw^{n-1}}{a}\, dw = \int F_1(w)\, dw,$$

where F_1 is a rational function if $F(\ ,\ )$ is rational.

Thus elementary functions suffice to handle even certain irrational functions, provided the worst irrationality is the nth root of a linear function. We can do still more, however. We can handle certain irrationalities involving quadratic functions, at least the square root of a quadratic. For example, if $F(\ ,\ )$ is a rational function of two variables we can integrate $\int F(z, \sqrt{1 - z^2})\, dz$ by letting $z = (w^2 - 1)/(w^2 + 1)$, so that $w = \sqrt{1 + z}/\sqrt{1 - z}$. Then

$$dz = 4w/(w^2 + 1)^2\, dw, \quad \sqrt{1 - z^2} = 2w/(w^2 + 1)$$

and so

$$\int F(z, \sqrt{1 - z^2})dz = \int F\left(\frac{w^2 - 1}{w + 1}, \frac{2w}{w^2 + 1}\right) \frac{4w}{(w^2 + 1)^2}\, dw$$

$$= \int F_1(w)\, dw,$$

where F_1 is again a rational function. For example, if this procedure is followed to evaluate $\int_0^w 1/\sqrt{1 - z^2}\, dz$, the result is $-i \log(\sqrt{1 - w^2} + iw)$. For the reader who has been taught to make the substitution $z = \sin\theta$ to evaluate integrals involving the square root of $1 - z^2$, and that the integral just evaluated is arcsin w, I would point out that the approach just described is equivalent to the substitution, because of the formula $\sin u = (e^{iu} - e^{-iu})/2i$, which implies that arcsin $w = -i \log(\sqrt{1 - w^2} + iw)$.

In the preceding discussion I have ignored the fact that the nth root is actually a multivalued function. Before any more integrals can profitably be discussed, this gap must be filled. We could take advantage of our previous work on the logarithm and define $z^{1/n}$ to be $e^{(1/n)\log z}$ for any acceptable value of the logarithm. If we did so, we would not find infinitely many different values for $z^{1/n}$. For the different values of log z differ by $2\pi k i$; hence the acceptable exponents in the expression for $z^{(1/n)}$ would differ by $2\pi k i/n$. But then values of k differing by a multiple of n would actually give the *same* value for $z^{(1/n)}$ because of the relation $e^{2\pi i} = 1$. Hence there are actually only n possible values for $z^{(1/n)}$. And indeed we find that although a possible value of $z^{(1/n)}$ does not return to its starting value after z traverses a circle around 0, it *does* return to the initial value when z traverses this circle n times. Thus the "spiral staircase" we imagined for the logarithm returns to the ground floor after ascending to floor $n - 1$. In terms of our Riemann surface, we should imagine the lower edge of the cut in the plane labeled $n - 1$ to be connected to the upper edge of the plane labeled 0 as in Figure A2-7. This connection is hard to make in terms of our edge-on picture, but we shall soon find a perfectly adequate way to visualize it.

The function $z^{(1/n)}$ is simpler than the logarithm in one other respect. It approaches a perfectly definite limit as z tends to 0, namely, 0. Also, the function $z^{(1/n)}$ is uniformly large for large values of z. These considerations suggest augmenting the complex plane by a "point at infinity" and assigning the value ∞ to certain functions where appropriate. (Where it is appropriate, the function will be said to have a *pole* at that point.) To get a picture of infinity which will simplify our Riemann surfaces, imagine the plane projected onto a sphere tangent to it by taking rays from the point antipodal

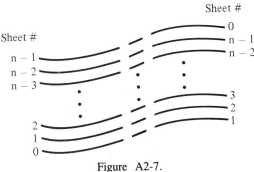

Figure A2-7.

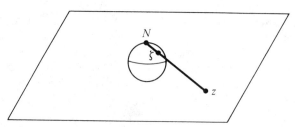

Figure A2-8.

to the point of tangency. The point from which the rays are drawn is labeled N in Figure A2-8. It corresponds to the value we shall call infinity. All other points of the sphere correspond to finite complex numbers.

   The cuts we made in the plane to define the Riemann surface of the logarithm and the function $z^{(1/n)}$ correspond to cuts in the sphere. But on the sphere it is very easy to see how to sew the cut edges together so that sphere $n - 1$ links up with sphere 0. Simply set the spheres down in a circle and sew each cut edge to the nearer cut edge on the sphere nearest that edge of the cut. In Figure A2-9, edges to be sewn together are labeled similarly. The spheres are arranged in a line rather than a circle for ease of sketching them. When all these spheres are glued together, the resulting Riemann surface still has the topological character of a sphere, that is, it could be stretched into a sphere without tearing it or sticking any two of its points together.

   Thus the Riemann surface of $z^{(1/n)}$ consists of a topological sphere. The same is true of the Riemann surface of a function involving the square root of a quadratic polynomial, e.g., $(1 - z^2)^{1/2}$. For this function, multi-valuedness occurs because the function does not return to its starting value when z describes a loop enclosing just one of the two points $+1$ and -1. It does return to its starting value, however, if the loop encloses *both* points ± 1. The easiest way to cut the plane so as to prevent z from traversing one of the undesirable loops is to cut out the line segment from -1 to $+1$. Since there are two admissible values for the square root, we simply take two copies of the plane (or sphere) with the line segments cut out, assign one of the values in a smooth (continuous) manner on each of the copies, then glue the edges together as before, to obtain one large "sphere" consisting of the two spheres glued together. It should be noted that even though $(1 - z^2)^{1/2}$ is single

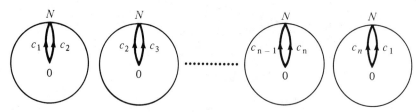

Figure A2-9.

valued on the Riemann surface, the integral $\int(1 - z^2)^{-1/2}dz = \arcsin z$ defines a multivalued function. The *inverse* function $\sin z$ is much better behaved, however. This point deserves to be looked at with the notation changed slightly. The function $(1 - z^2)^{1/2}$ can be thought of as the solution w to the equation $z^2 + w^2 - 1 = 0$, and the integral just written can then be written $\int(1/w)\, dz$. We have found once again that even when z is given a domain appropriate for eliminating the multivaluedness of w, the integral still defines a multivalued function. On the other hand, the inverse function to the integral is a well-behaved single-valued function. These principles can be formalized and generalized as follows:

(1)   For any polynomial equation $p(z, w) = 0$ a Riemann surface can be constructed so that the equation defines w as a single-valued function of z on the surface.
(2)   If $R(z, w)$ is a rational function of z and w, the equation

$$v = \int_a^u R(z, w)\, dz$$

defines v as a function of u. This function may be multivalued, i.e., the value of the integral may depend on the path chosen from a to u.
(3)   When v is a multivalued function of u, it is very likely that the inverse function will be a single-valued function. If so, it will be periodic.
(4)   If $p(x, y)$ is of degree two or less, the Riemann surface is a topological sphere and all such integrals are expressible using the elementary functions.

We shall now investigate some applications of these principles.

While the elementary functions suffice to solve many of the problems of classical mechanics, many of the most important and interesting differential equations lead to integrals which cannot be handled with elementary functions alone. For example, a simple model of pendulum motion leads to the equation

$$\frac{d^2\theta}{dt^2} + \omega^2 \sin \theta = 0,$$

where θ is the angle between the pendulum and the vertical at time r. To solve this equation let $\phi = d\theta/dt$, so that

$$\frac{d^2\theta}{dt^2} = \frac{d\theta}{dt} = \frac{d\phi}{d\theta}\frac{d\theta}{dt} = \phi\frac{d\phi}{d\theta}$$

The equation is thus

$$\phi\, d\phi = -\omega^2 \sin \theta\, d\theta.$$

Assuming an initial condition, say $\theta = \pi/10$ and $d\theta/dt = 0$ when $t = 0$, we find that if the values of θ and ϕ at time t_0 are, respectively, θ_0 and ϕ_0, then

$$\int_0^{\phi_0} \phi \, d\phi = \int_{\pi/10}^{\theta_0} -\omega^2 \sin \theta \, d\theta$$

so that

$$\frac{1}{2} \phi_0^2 = \omega^2 \left(\cos \theta_0 - \cos\left(\frac{1}{10}\pi\right) \right).$$

Since t_0 was an arbitrary time, we simply drop the subscript and find the following differential equation for θ:

$$\left(\frac{d\theta}{dt}\right)^2 = 2\omega^2 \left(\cos \theta - \cos \frac{1}{10}\pi \right).$$

If we let $x = \cos \theta$, $a = \cos(\pi/10)$, this equation becomes, upon integration,

$$t = \frac{-1}{\omega\sqrt{2}} \int_a^x \frac{dx}{\sqrt{(x-a)(1-x^2)}} \quad \text{or} \quad t = \frac{-1}{\omega\sqrt{2}} \int_a^x \frac{1}{y} \, dx,$$

where y is defined by the equation $y^2 - (x-a)(1-x^2) = 0$. Thus we have been led by a physical problem to consider the Riemann surface defined by a polynomial equation of degree 3. As it turns out, the integral just written cannot be expressed in terms of elementary functions. Our work on elementary functions has given us some idea what to expect, however. We expect that it will be easier to express x in terms of t than t in terms of x and that x will exhibit some periodicity. (Note that the physical model would lead us to expect both of these phenomena!) The conjecture is accurate, but it requires some auxiliary results to fill in the details.

We have just seen a good reason for studying the integral

$$\int \frac{dx}{\sqrt{(x-e_1)(x-e_2)(x-e_3)}} = \int \frac{dx}{\sqrt{P(x)}},$$

where the roots e_1, e_2, e_3 of $P(x)$ are real numbers. To do so we first construct the Riemann surface of the equation $y^2 - P(x) = 0$. Multivaluedness arises when x traverses loops which enclose an odd number of roots. To prevent that, we take two copies of the sphere (corresponding to the two possible values for the square root) and cut each copy from e_1 to e_2 and from e_3 to infinity. Our two spheres will then be sewn together along two different cuts, and the resulting surface will be a topological torus, that is, it will be deformable into the shape of an inner tube as in Figure A2-10.

It can be seen that this torus can be divided into four quarters, corresponding to the upper and lower half-planes in the two sheets of the Riemann surface. In order to study the mapping defined by the integral it is convenient to study these quarters one at a time. Accordingly let $w = \int_{z_o}^z (1/y) \, dx$, where z_o is any convenient point on the real axis less than the smallest root of the polynomial P. Imagine z tracing the contour in Figure A2-11.

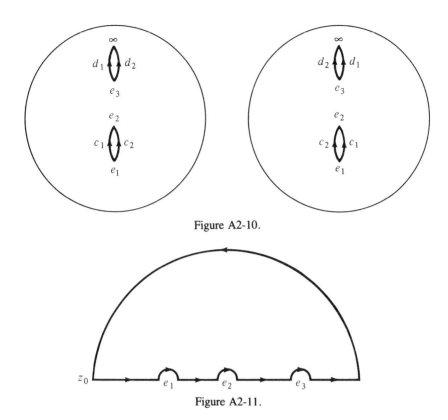

Figure A2-10.

Figure A2-11.

By noting that the square root is either real or pure imaginary and changes from real to pure imaginary and back again as z passes the roots e_1, e_2, e_3, we can see that the derivative of w along the real axis is alternately real and pure imaginary, hence that w is moving horizontally and vertically in alternation. Thus a qualitative picture of the image of the contour just shown is the contour of Figure A2-12, with corresponding points labeled.

If we shrink the small semicircles in Figure A2-11 to the points e_1, e_2, e_3 and expand to large contour to the point ∞, we see that the image of the upper half-plane is the rectangle whose corners are 0, iK', $-K + iK'$, $-K$ where

$$iK' = \int_{-\infty}^{e_1} \frac{1}{y}\,dx \qquad \text{and} \qquad K = -\int_{e_1}^{e_2} \frac{1}{y}\,dx.$$

Similar reasoning would show that the image of the lower half-plane in the same sheet is the rectangle whose corners are 0, iK', $K + iK'$, K. Since the values on the other sheet are the negatives of those on this sheet, the complete Riemann surface maps onto the larger rectangle in Figure A2-13. The integral we are studying is not single-valued on this Riemann surface, however. If z makes a loop which encloses e_1 and e_2 in sheet 1 but excludes e_3, then w will increase by $2K$, as can be seen in Figure A2-14.

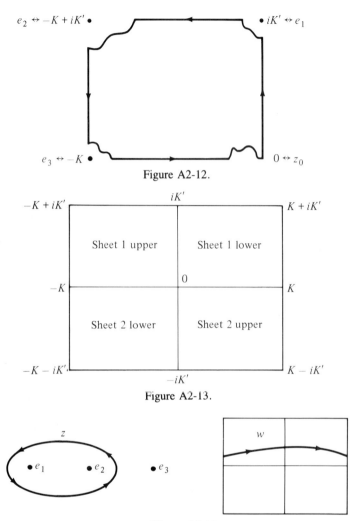

Figure A2-12.

Figure A2-13.

Figure A2-14.

Since z does not change its value when it returns to the starting point, but w moves from one side of the rectangle to the corresponding point on the opposite side, the opposite sides of the rectangle must be identified. Gluing the opposite edges of a rectangle together produces a torus, of course. The loop just described becomes a parallel of latitude on the torus if the edges it joins are connected last and a meridian of longitude if these edges are glued together first. The result of this identification is that the integral has essentially two independent loops (meridians of longitude on the torus and parallels of latitude) around which its value is not zero. The inverse function is therefore *doubly* periodic, and repeats its values if the plane is tiled with copies of the rectangle just drawn. In the case considered there are both real and imaginary periods.

Appendix 3.   Weierstrass' Formula

Let Σ be a subset of $R^3 - \{0\}$ such that every ray from $\mathbf{0}$ intersects Σ in precisely one point, for example,

$$\Sigma = \{\mathbf{x} : |x_1|^p + |x_2|^p + |x_3|^p = 1\}$$

where $\mathbf{x} = (x_1, x_2, x_3)$. (This convention will be adhered to throughout these appendices.) Then there is a unique function $\lambda : R^3 - \{0\} \to (0, \infty)$ such that for all $\mathbf{x}$, $\lambda^{-1}(\mathbf{x})\mathbf{x} \in \Sigma$. Hence Σ can be described as the surface whose equation is $\lambda(\mathbf{x}) = 1$. Obviously $\lambda(t\mathbf{x}) = t\lambda(\mathbf{x})$ for $t > 0$. If it happens that λ is continuously differentiable and grad $\lambda(\mathbf{x})$ never vanishes, then Σ is a differentiable manifold. It is easy to see that grad λ is constant on rays from $\mathbf{0}$.

The surface Σ can be parametrized by latitude and longitude:

$$\mathbf{v}(\phi, \theta) = \lambda^{-1}(\mathbf{x}(\phi, \theta))\mathbf{x}(\phi, \theta),$$

where $\mathbf{x}(\phi, \theta) = (\sin \phi \cos \theta, \sin \phi \sin \theta, \cos \phi)$ is a point on the unit sphere S^2. From this parametrization the surface area on Σ can be worked out. The computations use the fact that

$$\frac{\partial \mathbf{x}}{\partial \theta} \times \frac{\partial \mathbf{x}}{\partial \phi} = (\sin \phi)\mathbf{x}(\phi, \theta)$$

along with the identities $\boldsymbol{\xi} \cdot \text{grad } \lambda(\boldsymbol{\xi}) = \lambda(\boldsymbol{\xi})$ and grad $\lambda(t\boldsymbol{\xi}) = \text{grad } \lambda(\boldsymbol{\xi})$ and the well-known fact that $(\boldsymbol{\alpha} \times \boldsymbol{\beta}) \times \boldsymbol{\gamma} = (\boldsymbol{\alpha} \cdot \boldsymbol{\gamma})\boldsymbol{\beta} - (\boldsymbol{\beta} \cdot \boldsymbol{\gamma})\boldsymbol{\alpha}$. The final result is that an integral over Σ can be expressed as an integral over the sphere

$$\int_\Sigma f(\boldsymbol{\xi}) \, d\sigma(\boldsymbol{\xi}) = \int_{S^2} \lambda^{-2}(\boldsymbol{\eta})f(\lambda^{-1}(\boldsymbol{\eta})\boldsymbol{\eta}) |\text{grad } \lambda(\boldsymbol{\eta})| \, d\mu(\boldsymbol{\eta}). \tag{1}$$

In (1), $d\sigma$ and $d\mu$ are, respectively, the elements of surface area on Σ and S^2.

Integration over Σ is needed for two purposes in Weierstrass' method. First, it gives a formula for polar coordinates on R^3; second, it allows the application of the divergence theorem. The easiest way to obtain the polar coordinate formula is to introduce an auxiliary measure $d\tau$ on Σ whose relation to surface area can be worked out from equation (1). The measure $d\tau$ is defined as follows. For each subset E of Σ define $\tilde{E} = \{t\mathbf{x} : 0 < t \le 1, \mathbf{x} \in E\}$ and let $\tau(E) = 3m(\tilde{E})$, where m is Lebesgue measure on $R^3 - \{0\}$. It is obvious that τ is a Borel measure on Σ, and the formula

$$\int_\Sigma f(\boldsymbol{\xi}) \, d\tau(\boldsymbol{\xi}) = 3 \int_{\tilde{\Sigma}} f(\lambda^{-1}(\mathbf{x})\mathbf{x}) \, dm(\mathbf{x})$$

is an immediate consequence of the definition. The polar coordinate formula for τ

$$\int_{R^3 - \{0\}} g(\mathbf{x}) \, dm(\mathbf{x}) = \int_0^\infty \int_\Sigma g(t\boldsymbol{\xi}) \, d\tau(\boldsymbol{\xi})t^2 \, dt$$

follows immediately from the definition of τ if g is the characteristic function of a set of the form $\{t\mathbf{x} : \mathbf{x} \in E, \ a < t \le b\}$. Since every open set in $R^3 - \{0\}$ is a countable disjoint union of such sets, the equality holds for the characteristic function of every open set, and therefore for every continuous function g. It is easy to verify that for the special case of the surface $\Sigma = S^2$, so that $\lambda(\mathbf{x}) = |\mathbf{x}|$, the measure $d\tau$ is the same as $d\mu$, i.e., is surface area on the sphere. As a consequence we have a well-known polar coordinate formula

$$\int_{S^2} g(\xi)\, d\mu(\xi) = 3 \int_B g(|\mathbf{x}|^{-1}\mathbf{x})\, dm(\mathbf{x}).$$

where B is the ball $\{\mathbf{x} : 0 < |\mathbf{x}| \le 1\}$.

Formula (1) then gives

$$\int_\Sigma f(\xi)\, d\sigma(\xi) = 3 \int_B \lambda^{-3}(\mathbf{x}) |\mathbf{x}|^3 |\operatorname{grad} \lambda(\mathbf{x})| f(\lambda^{-1}(\mathbf{x})\mathbf{x})\, dm(\mathbf{x}).$$

In this last integral make the change of variable $\mathbf{y} = |\mathbf{x}|\lambda^{-1}(\mathbf{x})\mathbf{x}$. The Jacobian determinant for this transformation is $\lambda^{-3}(\mathbf{x})|\mathbf{x}|^3$, so that as a result

$$\int_\Sigma f(\xi)\, d\sigma(\xi) = 3 \int_{\tilde{\Sigma}} f(\lambda^{-1}(\mathbf{y})\mathbf{y}) |\operatorname{grad} \lambda(\mathbf{y})|\, dm(\mathbf{y})$$

$$= \int_\Sigma f(\xi) |\operatorname{grad} \lambda(\xi)|\, d\tau(\xi).$$

This formula gives the relation between the measures $d\tau$ and $d\sigma$ on Σ.

In view of the relation just established between $d\tau$ and $d\sigma$, the divergence theorem can be stated as

$$\int_\Sigma \mathbf{X}(\xi) \cdot \operatorname{grad} \lambda(\xi)\, d\tau(\xi) = \int_{\tilde{\Sigma} \cup \{0\}} \operatorname{div} \mathbf{X}(\mathbf{x})\, dm(\mathbf{x}).$$

Using this formula and the polar coordinate formula it is easy to establish Weierstrass' fundamental transformation

$$\frac{\partial}{\partial t} \left(\int_{\tilde{\Sigma}_t - \tilde{\Sigma}_{t_0}} D_i F(\mathbf{x})\, dm(\mathbf{x}) \right) = \frac{\partial^2}{\partial t^2} \left(\int_{\tilde{\Sigma}_t - \tilde{\Sigma}_{t_0}} F(\mathbf{x}) D_i \lambda(\mathbf{x})\, dm(\mathbf{x}) \right),$$

where $\tilde{\Sigma}_t = \{\mathbf{x} : 0 < \lambda(\mathbf{x}) \le t\}$. This formula is the basis for the method of solving certain partial differential equations described in Chapter 6.

Appendix 4.   Derivation of Euler's Equations

Consider two observers, one of whom is fixed in space and the other of whom is moving. If they each use a right-handed rectangular coordinate system to locate a point $\mathbf{P}$, then the coordinates X and Y which they assign to $\mathbf{P}$ will be related by an equation of the form

$$Y = A + RX$$

where

$$X = \begin{pmatrix} x_1 \\ x_2 \\ x_3 \end{pmatrix}, \qquad Y = \begin{pmatrix} y_1 \\ y_2 \\ y_3 \end{pmatrix}, \qquad A = \begin{pmatrix} a_1 \\ a_2 \\ a_3 \end{pmatrix},$$

and R is a 3-by-3 rotation matrix. This equation enables the observers to compare observations and decide whether they are observing the same point or not, i.e., two observations are the same vector **P** if the coordinates the observers assign to them are related as above.

Now that the two observers can agree on a way to compare positions, how are they to compare velocities? Clearly one should not expect that they will agree on the velocity of a moving point when they agree on its position; for they are moving with respect to each other. If they happen *not* to be moving with respect to each other, merely separated (by A) and differently oriented (by R), we should expect that they *will* agree on the velocity of the point. Now the coordinates of the velocity vectors as measured by the two observers are X' and Y'. Since A and R are constant when the two observers are not moving with respect to each other, the preceding equation gives $Y' = RX'$. Thus, in general, if the coordinates of position vectors are related by the equation $Y = A + RX$, then the coordinates of velocity vectors are related by $Z = RW$. This convention allows us to see how the two observers will disagree about velocities when they are moving with respect to each other. We have need only of the special case where $A = 0$, i.e., where the two observers have the same origin of coordinates. Henceforth we make that assumption. Then we have $Y(t) = R(t)X(t)$ and $Y' = RX' + R'X = R(X' + R*R'X)$, where $R* = R^{-1}$ is the transpose (or inverse) of R, since R is a rotation matrix. Assuming the first observer fixed in space, we shall say that his measurement of the velocity is the "true" one. For the moving observer, therefore, the "true" velocity of the point is RX'. However, the velocity that this observer *measures* is $Y' = RW$, where $W = X' + R*R'X$. Hence this observer will believe that the observer fixed in space has got his coordinates wrong by an amount $R*R'X$.

Now the fact that $R*R$ is the identity matrix, hence constant with respect to time, leads via simple differentiation, to the conclusion that $R*R'$ is skew-symmetric, i.e.,

$$R*R' = \begin{pmatrix} 0 & -c & b \\ c & 0 & -a \\ -b & a & 0 \end{pmatrix}.$$

If we identify this skew-symmetric matrix with the vector $(a, b, c) = \boldsymbol{\eta}$, we find that multiplying by the matrix corresponds to taking the cross product with $\boldsymbol{\eta}$. Thus when X gives the coordinates of the vector **P**, $R*R'X$ gives the coordinates of the vector $\boldsymbol{\eta} \times \mathbf{p}$. We can summarize all this by saying that for a moving observer the true velocity vector and the apparent velocity vector are related by the equation

$$\left(\frac{d\mathbf{P}}{dt}\right)_{\text{apparent}} = \left(\frac{d\mathbf{P}}{dt}\right)_{\text{true}} - \boldsymbol{\omega} \times \mathbf{P}.$$

The vector $\boldsymbol{\omega} = -\boldsymbol{\eta}$ is called the angular velocity of the rotating observer.

It is a simple matter, though not needed for what follows, to compare accelerations by repeating the argument just given. (Newton's second law will hold for the fixed observer, and the resulting equation shows that this law will hold for the rotating observer only if two "fictitious" forces called centrifugal force and Coriolis force are postulated.)

Now suppose point masses $m_1, \ldots, m_k$ located at points $\mathbf{r}_1(t), \ldots, \mathbf{r}_k(t)$ at time t are all rigidly attached to one another and to a fixed point which we take as 0. If a force $\mathbf{F}$ is applied at a point $\mathbf{r}$, also rigidly attached to the point masses and the fixed point, it will be transmitted to each of the point masses as a force $\mathbf{F}_i$. By Archimedes' law for levers we obtain

$$\mathbf{r} \times \mathbf{F} = \sum_{j=1}^{k} \mathbf{r}_j \times \mathbf{F}_j = \sum_{j=1}^{k} m_j\left(\mathbf{r}_j \times \frac{d^2}{dt^2}\mathbf{r}_j\right).$$

Since the masses are rigidly attached to one another and to the origin, their motion can only be a rotation, and in a nonmoving coordinate system we have $d\mathbf{r}_j/dt = \boldsymbol{\omega} \times \mathbf{r}_j$ for some vector $\boldsymbol{\omega}$, the same for all j. (For, in a frame of reference fixed with respect to the masses m_j we obviously have $d\mathbf{r}_j/dt = \mathbf{0}$.) Thus in spatial (nonmoving) coordinates we have

$$\mathbf{r} \times \mathbf{F} = \sum_{j=1}^{k} m_j\left(\mathbf{r}_j \times \frac{d}{dt}(\boldsymbol{\omega} \times \mathbf{r}_j)\right),$$

and in a frame of reference attached to the masses m_j

$$\mathbf{r} \times \mathbf{F} = \sum_{j=1}^{k} m_j\left(\mathbf{r}_j \times \frac{d}{dt}(\boldsymbol{\omega} \times \mathbf{r}_j) + \mathbf{r}_j \times (\boldsymbol{\omega} \times (\boldsymbol{\omega} \times \mathbf{r}_j))\right)$$

$$= \sum_{j=1}^{k} m_j\left(\frac{d}{dt}(\mathbf{r}_j \times (\boldsymbol{\omega} \times \mathbf{r}_j)) - \mathbf{r}_j \times (\boldsymbol{\omega} \times (\mathbf{r}_j \times \boldsymbol{\omega}))\right).$$

If we then use the trivial identity $\mathbf{a} \times (\mathbf{b} \times (\mathbf{a} \times \mathbf{b})) = -\mathbf{b} \times (\mathbf{a} \times (\mathbf{b} \times \mathbf{a}))$, we find

$$\mathbf{r} \times \mathbf{F} = \frac{d}{dt}(T\boldsymbol{\omega}) + \boldsymbol{\omega} \times (T\boldsymbol{\omega}),$$

where $T = \sum_{j=1}^{k} m_j\mathbf{r}_j \times (\boldsymbol{\omega} \times \mathbf{r}_j)$ is called the inertia tensor for the system.

For a continuous body with density $\rho(\mathbf{x})$ the usual limiting argument shows that the same equation holds with the inertia tensor being defined as

$$T\boldsymbol{\omega} = \int \rho(\mathbf{x})\mathbf{x} \times (\boldsymbol{\omega} \times \mathbf{x}) \, dV(\mathbf{x})$$

and the volume integral being extended over the region occupied by the body. Simply by calculating the principal minors of a matrix representing the inertia tensor in an orthonormal coordinate system, one can see that the inertia tensor is a symmetric, positive-definite operator. Therefore in some orthonormal coordinate system, called the system of principal axes of the body, this operator has a diagonal matrix. (This is the reason for stating the Euler equations in terms of body coordinates rather than spatial coordinates.)

If the only force acting on the body is gravitational, the force may be taken as $Mg\boldsymbol{\gamma}$, where $\boldsymbol{\gamma}$ is a unit vector pointing downward, and the force may be considered as attached at the center of gravity $\mathbf{X}$. Then we obtain from what has been said above

$$Mg\mathbf{X} \times \boldsymbol{\gamma} = T\left(\frac{d\boldsymbol{\omega}}{dt}\right) + \boldsymbol{\omega} \times (T\boldsymbol{\omega}).$$

Since the vector $\boldsymbol{\gamma}$ is constant in a spatial frame of reference, we find that in the rotating frame of reference

$$\frac{d\boldsymbol{\gamma}}{dt} + \boldsymbol{\omega} \times \boldsymbol{\gamma} = \mathbf{0}.$$

Appendix 5.   Calculus of Variations

Suppose we wish to find a function $y(x)$ which gives a minimum or maximum value to the integral

$$\int_a^b f(x, y(x), y'(x))\, dx$$

and satisfies $y(a) = A$ and $y(b) = B$. Assuming there is such a function $y(x)$, change it slightly by forming the function $y(x) + tz(x)$, where t is small and $z(a) = z(b) = 0$. The function

$$\phi(t) = \int_a^b f(x, y(x) + tz(x)y'(x) + tz'(x))\, dx$$

will have an extreme value at $t = 0$. Hence we will have $\phi'(0) = 0$. But this means

$$\int_a^b D_2 f(x, y(x), y'(x))z(x) + D_3 f(x, y(x)), y'(x))z'(x)\, dx = 0.$$

If we integrate by parts in the second term, taking $u = D_3 f(x, y(x), y'(x))$ and $dv = z'(x)\, dx$, we shall have

$$0 = \int_a^b (D_2 f(x, y(x), y'(x)) - \frac{d}{dx} D_3 f(x, y(x), y'(x)))z(x)\, dx.$$

Since $z(x)$ is arbitrary, it is not difficult to show that this equation implies that the integrand is identically zero, i.e., that

$$D_2 f(x, y(x), y'(x)) = \frac{d}{dx} D_3 f(x, y(x), y'(x)).$$

This last equation is known as Euler's equation in calculus of variations. In classical notation it was written

$$\frac{\partial f}{\partial y} = \frac{d}{dx}\left(\frac{\partial f}{\partial y'}\right).$$

Appendix 6. Jacobi's Last-Multiplier Method

In the nineteenth century solving a first-order, first-degree differential equation

$$\frac{dx}{X(x, y)} = \frac{dy}{Y(x, y)}$$

was taken to mean finding an "integral," i.e., a nonconstant function $F(x, y)$ which is constant when x and y vary so as to satisfy the equation. For our present purposes we may modernize this notation and say that an integral is simply a function whose gradient is perpendicular to the vector (X, Y), i.e., $X D_1 F + Y D_2 F = 0$. This is easy to do if the equation is exact, i.e., $\mathrm{div}(X, Y) = D_1 X + D_2 Y = 0$. For then we know we can find a function f such that $D_1 f = -Y$ and $D_2 f = X$, and then f is the required integral. The method for finding f is well known and involves only being able to integrate. More generally this method applies if an integrating factor can be found so that $\mathrm{div}(mX, mY) = 0$.

Jacobi's method applies in the more general case of a system of $n - 1$ equations in n variables, i.e.,

$$\frac{dx_1}{X_1} = \frac{dx_2}{X_2} = \cdots = \frac{dx_n}{X_n}$$

provided two conditions are met. First $\mathrm{div}(X_1, \ldots, X_n) = D_1 X_1 + \cdots + D_n X_n = 0$ [or more generally a multiplier can be found such that this condition holds with $(mX_1, \ldots, mX_n)$] and, second, one has already found $n - 2$ independent integrals, say $F_3(X_1, \ldots, X_n), \ldots, F_n(X_1, \ldots, X_n)$. Given these hypotheses, the last integral $F_2(X_1, \ldots, X_n)$ can be found by the following procedure.

Independence of the integrals already found is taken to mean that their Jacobian matrix has rank $n - 2$. Hence there is no loss in generality in assuming that the matrix

$$M = \begin{pmatrix} D_3 F_3 & \cdots & D_n F_3 \\ \vdots & & \vdots \\ D_3 F_n & \cdots & D_n F_n \end{pmatrix}$$

is invertible. In that case the system of equations

$$y_1 = x_1,$$
$$y_2 = x_2,$$
$$y_3 = F_3(x_1, \ldots, x_n),$$
$$\vdots$$
$$y_n = F_n(x_1, \ldots, x_n)$$

can be solved for the x's in terms of the y's giving

$$x_1 = y_1,$$
$$x_2 = y_2,$$
$$x_3 = G_3(y_1, \ldots, y_n),$$
$$\vdots$$
$$x_n = G_n(y_1, \ldots, y_n).$$

Let $N(y_1, \ldots, y_n)$ denote the Jacobian of this last transformation, so that at corresponding points

$$N(y_1, \ldots, y_n) = \frac{1}{\det M(x_1, \ldots, x_n)}$$

and set

$$Y_j(y_1, \ldots, y_n)$$
$$= N(y_1, \ldots, y_n)X_j(y_1, y_2, G_3(y_1, \ldots, y_n), \ldots, G_n(y_1, \ldots, y_n)).$$

By very tedious computation one can verify that $D_1 Y_1 + D_2 Y_2 = 0$, so that the equation

$$\frac{dy_1}{Y_1} = \frac{dy_2}{Y_2}$$

can be solved by quadrature for fixed but arbitrary values of $y_3, \ldots, y_n$. (The function N is the last multiplier which gives this method its name, since it makes the equation exact when it is multiplied by X_1 and X_2.) Then we can find a function $f(y_1, \ldots, y_n)$ such that $D_1 f = -Y_2$ and $D_2 f = Y_1$. Another tedious but entirely routine computation reveals that the function $F_2(x_1, \ldots, x_n)$ given by

$$F_2(x_1, \ldots, x_n) = f(x_1, x_2, F_3(x_1, \ldots, x_n), \ldots, F_n(x_1, \ldots, x_n))$$

is a new integral independent of $F_3, \ldots, F_n$, and in fact that

$$(X_1, X_2, \ldots, X_n) = \operatorname{grad} F_2 \times \operatorname{grad} F_3 \times \cdots \times \operatorname{grad} F_n$$

$$= \det \begin{pmatrix} \mathbf{e}_1 & \mathbf{e}_2 & \cdots & \mathbf{e}_n \\ D_1 F_2 & D_2 F_2 & \cdots & D_n F_2 \\ D_1 F_n & D_2 F_n & \cdots & D_n F_n \end{pmatrix},$$

where $e_1, \ldots, e_n$ is the standard basis of R^n.

As an example for practice the reader may take the equations

$$\frac{dx}{ry - qz} = \frac{dy}{pz - rx} = \frac{dz}{qx - py}$$

regarding p, q, and r as constants, and the integral $F_3(x, y, z) = x^2 + y^2 + z^2$. Then if $u = x$, $v = y$, $w = x^2 + y^2 + z^2$, we find that the last multiplier is $N(u, v, w) = \frac{1}{2}(w - u^2 - v^2)^{-1/2}$, and the equation

$$\frac{du}{\frac{1}{2}rv(w - u - v)^{-1/2} - \frac{1}{2}q} = \frac{dv}{\frac{1}{2}p - \frac{1}{2}ru(w - u - v)^{-1/2}}$$

is exact. In fact the function $f(u, v, w) = -\frac{1}{2}(pu + qv + r(w - u^2 - v^2)^{-1/2})$ has the properties required for the argument above, and we find as a second integral the function $F_2(x, y, z) = -\frac{1}{2}(px + qy + rz)$. It is easy to verify that grad $F_2 \times$ grad $F_3 = (ry - qz, pz - rx, qx - py)$. .

Bibliography

In this bibliography the following abbreviations will be used: *JFM* = *Journal für die reine und angewandte Mathematik; CR* = *Comptes Rendus de l'Académie des Sciences; AM* = *Acta Mathematica*.

ABEL, N. H.

1826   Untersuchungen über die Reihe $1 + (m/1)x + (m(m-1)/2)x^2 + \cdots$. *JFM* 1; 311–339. French translation in *Oeuvres* I; 219–250.
1827–28   Recherches sur les fonctions élliptiques. *JFM* 2–4. *Oeuvres* I; 263–388.
1829   Remarques sur quelques propriétés générales d'une certaine sorte de fonctions transcendantes. *JFM* 3; 313–323 = *Oeuvres* I; 444–456.
          Oeuvres Complètes. Two volumes. Christiania: Grondahl & Son, 1881. Johnson Reprint Corporation, New York, 1965.

APPEL'ROT, G. G.

1892   On paragraph 1 of S. V. Kovalevskaya's memoir 'Sur le problème de rotation d'un corps solide autour d'un point fixe' (Russian). *Matematicheskii Sbornik* T. 16, No. 3; 487–507.

BAKER, H. F.

1897   *Abel's Theorem and the Allied Theory Including the Theory of the Theta Functions*. Cambridge University Press.

BEHNKE, H. and KOPFERMANN, K., EDS.

1966   *Festschrift zur Gedächtnisfeier für Karl Weierstrass 1815–1965*. Köln und Opladen: Westdeutscher Verlag.

BELL, E. T.

1937   *Men of Mathematics*. New York: Simon and Schuster.

BERNSTEIN, S.

1908   The analytic nature of solutions of differential equations of elliptic type (Russian). Master's thesis. Reprint by Kharkov State University Press, 1956.

BESSEL, F. W.

1807   Über die Figur des Saturns, mit Rücksicht auf die Attraction seiner Ringe. *Monatliche Correspondenz zur Beförderung der Erd- und Himmels-Kunde* XV; 239–260.
1812   Untersuchungen über den Planeten Saturn, seinen Ring und seinen vierten Trabanten. *Königsberger Archives* I; 113–172.

BIERMANN, K.-R.

1965   Die Berufung von Weierstrass nach Berlin. In Behnke and Kopfermann, eds., 1966; 41–52.
1976   Weierstrass. In *Dictionary of Scientific Biography* XIV; 219–224.

BLISS, G. A.

1925   *Calculus of Variations*. Mathematical Association of America, Carus Monograph No. 1. Open Court Publishing Co., LaSalle, Illinois.
1933   *Algebraic Functions*. American Mathematical Society Colloquium Publication XVI. Dover, New York, 1966.

DU BOIS-REYMOND, P.

1875   Versuch einer Classification der willkürlichen Functionen. *JFM* 79; 21–37.
1879   Erläuterungen zu den Anfangsgründen der Variationsrechnung. *Mathematische Annalen* 15; 283–314, 564–576.
1882   *Die allgemeine Funktionentheorie*. Tübingen: H. Laupp.

BOOLE, G.

1859   *A Treatise on Differential Equations*. Fifth edition, revised by I. Todhunter, 1865. Chelsea Publishing Co., New York

BOWDITCH, N.

1829–39   *Mécanique Céleste of Laplace*, translated by N. Bowditch. Boston: Hillard, Gray, Little & Wilkins.

BOWMAN, R.

1953   *Introduction to Elliptic Functions, with Applications*. London: English Universities Press, Ltd.

BRIOT, Ch. and BOUQUET, J.-K.

1856   Recherches sur les fonctions définies par des équations différentielles. *Journal de l'École Polytechnique* T. XXI, Cahier 36; 133–198.

BRILL, A. and NOETHER, M.

1894   Die Entwicklung der Theorie der algebraischen Functionen in älterer und neuerer Zeit. *Jahresbericht der deutschen Mathematiker-Vereiningung* 3; 107–566.

BRUNS, E. H.

1871   *De proprietate quadam functionis potentialis corporum homogeneorum*. Inaugural dissertation, Berlin.
1887   Über die Integrale des Vielkörperproblemes. *AM* 11; 25–96.
        Cahiers du Séminaire d'Histoire des Mathématiques 4; 75–87(1983). Partie inédite de la correspondance de Hermite avec Stieltjes. Paris: Institute Henri Poincaré.

CAUCHY, A. L.

1814   Mémoire sur la théorie de la propagation des ondes à la surface d'un fluide pesant d'une profondeur indéfinie. *Mémoires des Savants Étrangers* I; 1827, 3–312.
1826–30   *Exercises de Mathématiques*. Paris: De Bure frères.
1840–47   *Exercises d'Analyse et de Physique Mathématique*. Paris: Bachelier.
1842a   Mémoire sur un théorème fondamental dans le calcul intégrale. *CR* XIV; 1020–1026.
1842b   Mémoire sur l'emploi du nouveau calcul des limites dans l'intégration d'un système d'équations différentielles. *CR* XV; 14–25.
1842c   Mémoire sur l'emploi du calcul des limites dans l'intégration des équations aux dérivées partielles. *CR* XV; 44–59.
1842d   Mémoire sur l'application du calcul des limites à l'intégration d'un système d'équations aux dérivées partielles. *CR* XV; 85–101.
1842e   Mémoire sur les systèmes d'équations aux dérivées partielles d'ordre quelconque, et sur leur réduction à des systèmes d'équations linéaires du premier ordre. *CR* XV; 131–138.

COHN, P. M.

1957   *Lie Groups*. Cambridge Tracts in Mathematical Physics No. 46. Cambridge University Press.

DARBOUX, G.

1875a   Mémoire sur l'éxistence de l'intégrale dans les équations aux dérivées
         partielles contenant un nombre quelconque de fonctions et de variables
         indépendants. *CR* LXXX; 101–104.
1875b   Sur l'éxistence de l'intégrale dans les équations aux dérivées partielles
         d'ordre quelconque. *CR* LXXX; 317–319.

DOMAR, Y.

1982   On the foundation of *Acta Mathematica*. *AM* 148; 3–8.

DUBROVIN, B. A., NOVIKOV, S. P., and MATVEEV, V. B.

1976   Nonlinear equations of Korteweg-deVries type. *Uspekhi Mat-
         ematicheskikh Nauk* 31; 55–136 (Russian).

ENGELMANN, W., ed.

1880   *Briefwechsel zwischen Gauss und Bessel*. Königliche Preussische Aka-
         demie der Wissenschaften, Leipzig.

ERICKSON, Å.

1981   Matematikforskning på internationell nivå. *Värt att veta i Danderyd*
         (local publication in Danderyd, Sweden). Juni 1981, Nr. 3; 10–11.

EULER, L.

1736   *Mechanica, sive Motus scientia analytice exposita*. Petersburg Acad-
         emy of Sciences.
1748   *Introductio in analysin infinitorum*. Lausanne: M. M. Bousquet.
1752   De integratione aequationis differentialis $m\,dx/\sqrt{1-x^4} =
         n\,dx/\sqrt{1-y^4}$. *Novi Comentarii Acad. Petropolitanae* T. 6; 37.
1758   Du mouvement de rotation des corps solides autour d'un axe variable.
         Mémoires de l'Académie des Sciences de Berlin XIV; 154–193.
1765   *Theoria Motus Corporum Solidorum seu Rigidorum*. Greifswald: Ros-
         tock. *Opera*, 2nd ser. III, IV.

FORSTER, O.

1981   *Lectures on Riemann Surfaces*. Translated from the German by Bruce
         Gilligan. New York: Springer–Verlag.

FORSYTH, A. R.

1918   *Theory of Functions of a Complex Variable.* Cambridge University
        Press.

FREDHOLM, I.

1890   Om en speciell klass af singulära linjer. *Öfversigt af Kung. Vetenskaps-
        Akademiens Förhandlingar* 1890; 131–134.

FRESNEL, A. J.

1866–1870   *Oeuvres Complètes.* Paris: Imprimerie Impériale. 3 Vols.

FRICKE, R.

1913   Elliptische Funktionen. In *Enzyklopädie der Mathematischen Wis-
        senschaften* Band II. 2 Teil; 177–345. Leipzig: Teubner.

FREUDENTHAL, H.

1971   Cauchy. In *Dictionary of Scientific Biography* III; 131–148.

FROSTMAN, O.

1966   Aus dem Briefwechsel von G. Mittag-Leffler. In Behnke and Kop-
        fermann, eds., 1966; 53–56.

FUCHS, L.

1865   Zur Theorie der linearen Differentialgleichungen mit veränderlichen
        Koeffizienten. *Jahresbericht über die städtische Gewerbeschule zu Ber-
        lin.*
1884   Über Differentialgleichungen deren Integrale fixte Ver-
        zweigungspunkten besitzen. *Monatsberichte der Akademie der Wis-
        senschaften zu Berlin.*

GAUSS, K. F. W.

        Gesammelte Werke. 12 Vols. Königliche Gesellschaft der
        Wissenschaften zu Göttingen. Leipzig-Berlin, 1863–1933.

GELFAND, I. M.

1941   Normierte Ringe. *Matematicheskii Sbornik* 9; 3–24.

GENNOCHI, A.

1875   Observations relatives à une communication précedente de M. Dar-
            boux. *CR LXXX*; 315–316.

GÖPEL, A.

1847   Theoriae transcendentium Abelianarum primi ordinis adumbratio levis.
            JFM 35; 277–312.

GOLUBEV, V. V.

1950   The works of S. V. Kovalevskaya on the motion of a rigid body about
            a fixed point (Russian) Institute of Mechanics of the USSR Academy of
            Sciences. *Applied Mathematics and Mechanics* 14; 236–244.

1953   *Lectures on integration of the equations of motion of a rigid body about
            a fixed point.* Translated from the Russian by J. Shorr-Kon. Israel
            Program for Scientific Translation. National Science Foundation.

GRATTAN-GUINNESS, I.

1971   Materials for the history of mathematics in the Institut Mittag-Leffler.
            Isis vol. 62, 3, No. 231; 363–374.

1972   A Mathematical Union: William Henry and Grace Chisholm Young.
            Annals of Science 29, No. 2; 105–186.

1981   Mathematical Physics in France, 1800–1840: Knowledge, Activity, and
            Historiography. In *Mathematical Perspectives,* J. Dauben, ed. New
            York: Academic Press; 95–138.

1982   Unpublished manuscript. Chapter 9, The entry of Cauchy: Complex
            variables and differential equations, 1810–1822. Chapter 10, The inau-
            guration of mathematical analysis, 1820–1826.

GRAY, J.

1984   Fuchs and the theory of differential equations. *Bulletin of the American
            Mathematical Society* Volume 10, No. 1; 1–26.

HADAMARD, J.

1895   Sur la stabilité des rotations dans un mouvement d'un corps solide
            pesant autour d'un point fixe. Paper read 5 August 1895 at a meeting of
            the Association Française at Bordeaux = *Oeuvres* IV; 1719–1724.

1923   *Lectures on Cauchy's Problem in Linear Partial Differential Equations.*
         Yale University Press. Dover, New York, 1952.
1935   Équations aux dérivées partielles. Paper read at the Conférence Inter-
         nationale sur les Equations aux Dérivées Partielles, 17–20 June
         1935 = *Oeuvres* III; 1593–1630.
         Oeuvres. 4 Vols. Centre Nationale de la Recherche Scientifique. Paris,
         1968.

HAMMERSTEIN, A.

1930   Nichtlineare Integralgleichungen nebst Anwendungen. *AM* 54;
         117–176.

HAWKINS, T.

1977   Weierstrass and the theory of matrices. *Archives for the History of
         Science* 17, No. 2; 119–163.

HAYWARD, R. B.

1858   On a direct method of estimating velocities, accelerations, and all sim-
         ilar quantities with respect to axes movable in any manner in space.
         Transactions of the Cambridge Philosophical Society X; 1–22.

HERMITE, Ch.

1844   Sur la théorie des transcendantes à différentielles algébriques.
         Liouville's Journal IX; 353–368 = *Oeuvres* I; 49–63.
1848   Sur la division des fonctions abéliennes ou ultra-élliptiques. *Mémoires
         Présentés par Divers Savants* X; 563–572 = *Oeuvres* I; 38–48.
1855   Sur la théorie de la transformation des fonctions abéliennes. *CR* XI;
         249–254, 304–309, 365–369, 427–431, 485–489, 536–541, 704–707,
         784–787 = *Oeuvres* I; 444–478.
         Oeuvres. E. Picard ed., 4 vols. Paris: Gauthier-Villars, 1905–1917.

HILBERT, D.

1901   Über das Dirichletsche Prinzip. Address at 150th anniversary of the
         Königliche Gesellschaft der Wissenschaften zu Göttingen = *Mathe-
         matische Annalen* 59 (1904); 161–186 = *Abhandlungen* III; 15–37.
         Gesammelte Abhandlungen. 3 vols. Berlin, 1932–1935. Chelsea, New
         York, 1965.

HILLE, E.

1962   In retrospect. Address at Yale University 16 May 1962. *Mathematical
         Intelligencer* Vol. 3, No. 1 (1980); 3–13.

HUYGHENS, Ch.

1690   *Traité de la lumière*. Leyden: P. van der Aa.

JACOBI, C. G. J.

1828   Sur les fonctions élliptiques. *JFM* 3; 192–196, 303–310, 403–404.
1829   *Fundamenta nova theoriae functionum ellipticarum.* Königsberg: Born-
            traeger.
1832a   De theoremate Abeliano observatio. *JFM* 9; 99.
1832b   Considerationes generales de transcendentibus Abelianis. *JFM* 9;
            394–403.
1832c   Über und zu Legendre's 'Théorie des fonctions élliptiques.' *JFM* 8;
            413–418.
1835   De functionibus duarum variabilium quadruplicitis periodicis quibum
            theoria transcendentium Abelianarum innititus. *JFM* 13; 55–78.
1849   Sur la rotation d'un corps. *CR* XXIX; 97–106. *JFM* 39; 293–350.
            Liouville's Journal XIV; 337–344.
1862   Nova methodus aequationes differentiales partiales primi ordinis inter
            numerum variabilium quamcunque propositas integrandi. *JFM* 60;
            1–181.
1865   De investigando ordine systematis differentialium vulgarium cujuscun-
            que. *JFM* 64; 297–320.
            De aequationum differentialium systemate non normali ad formam nor-
            malem revocando auctore. In *Werke* 5; 483–513.
            Gesammelte Werke. 7 vols. Berlin, 1881–1891.

JOUKOWSKY, N. see ZHUKOVSKY, N.

KLEIN, F.

1897   *The Mathematical Theory of the Top*. New York: Charles Scribner's
            Sons.
1926   *Vorlesungen über die Entwicklung der Mathematik im 19. Jahrhundert.*
            Berlin: Springer-Verlag. Chelsea, New York, 1956. 2 vols., bound
            as one.

KLEIN, F. and SOMMERFELD, P.

1897–1910 *Theorie des Kreisels.* 4 vols. Leipzig: Teubner.

KLINE, M.

1972   *Mathematical Thought From Ancient to Modern Times*. New York:
            Oxford University Press.

KOBLITZ, A. H.

1983   *A Convergence of Lives. Sophia Kovalevskaia: Scientist, Writer, Revolutionary*. Boston: Birkhäuser.

KOCHINA, P. Ya.

1973   *Briefe von Weierstrass an Sophie Kowalewskaja*. Moscow: "Nauka."

1974   *S. V. Kovalevskaya. Reminiscences and Stories* (Russian): Moscow: Akademiia Nauk, USSR.

1979   *Letters to S. V. Kovalevskaya From Foreign Mathematicians*. (Russian). (Preprint) Institute for the Problems of Mechanics of the USSR Academy of Sciences, No. 121.

1980   A letter of H. A. Schwarz to S. V. Kovalevskaya (Russian) *Voprosy Istorii Estestvoznania i Tekhniki* 7; 105–111.

1981   *Sof'ya Vasilievna Kovalevskaya*. (Russian) Moscow: "Nauka."

KOENIGSBERGER, L.

1865   Über die Transformationen der abel'schen Functionen erster Ordnung. *JFM* 64; 17–42, *JFM* 65; 335–358.

1867   Über die Transformation des zweiten Grades für die abelschen Functionen erster Ordnung. *JFM* 67; 58–77.

1906   *Hermann von Helmholtz*. English translation by Frances A. Welby. Oxford: Clarendon Press. Dover, New York, 1965.

KORVIN-KRUKOVSKY, F. V.

1891   *Sophia Vasilievna Kovalevskaya, née Korvin-Krukovskaya* (Russian). *Russkaya Starina* 71, No. 9; 623–636 = *Shtraikh* 1951; 370–384.

KOVALEVSKAYA, S. V.

1875   Zur Theorie der partiellen Differentialgleichungen. *JFM* 80; 1–32. Russian translation in *Raboty*, 7–50.

1884   Über die Reduction einer bestimmten Klasse abel'scher Integrale dritten Ranges auf elliptische Integrale. *AM* 4; 393–414. Russian translation in *Raboty*, 51–74.

1885a   Über die Brechung des Lichtes in crystallinischen Mitteln. *AM* 6; 249–304. Russian translation in *Raboty*, 75–138. French résumé in *CR* 98; 356–357. Swedish résumé in *Öfversigt af Kungl. Vetenskaps-Akademiens Förhandlingar* 41; 119–121.

1885b   Zusätze und Bemerkungen zu Laplace's Untersuchung über die Gestalt der Saturnringe. *Astronomische Nachrichten* 111; 37–48. Russian translation in *Raboty*, 139–152.

1886   Reminiscences of George Eliot (Russian). *Russkaya Mysl* 6; 93–108.

1889   Mémoire sur un cas particulier de la rotation d'un corps solide autour
           d'un point fixe. *AM* 12; 177–232. Russian translation of paragraphs 1–4
           and 9 in *Raboty*, 235–244. The rest of the article is identical to 1890a.

1890a   Mémoire sur un cas particulier du problème de la rotation d'un corps
           solide autour d'un point fixe, où l'intégration s'éffectue à l'aide de
           fonctions ultraélliptiques du temps. *Mémoires Présentés par Divers
           Savants* 31; 1–62. Russian translation in *Raboty*, 153–220.

1890b   Sur une propriété du système d'équations différentielles qui définit la
           rotation d'un corps solide autour d'un point fixe. *AM* 14; 81–93. Rus-
           sian translation in *Raboty*, 221–234.

1890c   An autobiographical sketch (Russian). *Russkaya Starina* 11; 450–463.
           English translation in *Stillman* 1978, 213–229.

1890d   *Memories of Childhood* (Russian). *Vestnik Evropy* 7; 55–98, 8;
           584–640. English translation in *Leffler 1895* and in *Stillman 1978*.

1891   Sur un théorème de M. Bruns. *AM* 15; 45–52. Russian translation in
           Raboty, 245–254.

1892   *The Nihilist Woman* (Russian). Geneva: Vol'naya Russkaya Tipografia.

Raboty   *S. V. Kovalevskaya. Scientific Works* (Russian). Moscow: USSR Acad-
           emy of Sciences, 1948.

Nachlass   References to Kovalevskaya's unpublished material in the archives of
           the USSR Academy of Sciences, 1948, can be found in *Kochina 1981*
           and *Koblitz 1983*. The material in the Institut Mittag-Leffler was de-
           scribed in *Grattan-Guinness 1971*.

KOVALEVSKY, M. M.

1889   *Modern Customs and Ancient Laws of Russia, Being the Ilchester Lec-
           tures for 1889–1890*. London: David Nutt, 1891.

KRAMER, E.

1973   Sonya Kovalevsky. *Dictionary of Scientific Biography* VII; 477–479.

KRAZER, A.

1903   *Lehrbuch der Thetafunktionen*. Leipzig. Chelsea, New York, 1970.

KRAZER, A. and WIRTINGER, W.

1921   Abelsche Funktionen and allgemeine Thetafunktionen. *Enzyklopädie
           der Mathematischen Wissenschaften* II, Band 2; 604–873.

LAGRANGE, J. L.

1811–1815   *Mechanique Analytique, Nouvelle édition*. 2 vols. Paris.

LAMÉ, G.

1866   *Leçons sur la Théorie de l'Élasticité des Corps Solides, Deuxième Édition.* Paris: Gauthier-Villars.

LAPLACE, P. S.

1799   *Traité de Mécanique Céleste.* Tome 2. Paris: Duprat, an VII (1799).

LEFFLER, A.-Ch.

1895   *Sonya Kovalevsky.* Her Recollections of Childhood with a biography by Anna Carlotta Leffler, Duchess of Cajanello. Translated by Isabel F. Hapgood and A. M. Clive Bayley. New York: The Century Company.

LEGENDRE, A. M.

1825–1828   *Traité des Fonctions Élliptiques et des Intégrales Euleriennes.*

LERMONTOVA, Ju. V.

Reminiscences of Sophia Kovalevskaya (Russian). In *Shtraikh* 1951; 375–387.

LIOUVILLE, R.

1897   Sur le mouvement d'un corps solide pesant suspendu par l'un de ses points. *AM* 20; 239–284.

LITVINOVA, E. F.

1894   *S. V. Kovalevskaya, Woman Mathematician, Her Life and Scholarly Work. A Biographical Essay* (Russian). St. Petersburg: Pavlenkov.

1897   *Rulers and Thinkers. Bibliographical Essays* (Russian). St. Petersburg: Pavlenkov.

LYAPUNOV, A. M.

1894   On a property of the differential equations for the problem of the motion of a rigid solid body having a fixed point (Russian). *Soobshchenia Kharkov. Matematich. Obshchestva,* 2nd series, 4; 123–140.

MALEVICH, J.

1890   Sofia Vasilievna Kovalevskaya, doctor of philosophy and professor of higher mathematics, in "Reminiscences of her chronologically first teacher J. Malevich" (Russian). *Russkaya Starina* 12; 615–654.

MANNING, K. P.

1975   The emergence of the Weierstrassian approach to complex analysis. *Archives for the History of the Exact Sciences* 14, No. 4; 297–383.

MAXWELL, J.

1859   On the stability of the motion of Saturn's rings. *Astronomical Society of London, Monthly Notices* XIX; 297–304. See also *Niven* I; 288 = 376.

MENDEL'SON, M.

1912   Reminiscences of Sophia Kovalevskaya, with her letters. Russian translation (from the Polish) by L. Krukovskaya. *Sovremenny Mir* No. 2; 134–176.

MITTAG-LEFFLER, G.

1876   En metod att analytiskt framställa en funktion af rationel karakter hvilker blir oändlig alltid och endast uti vissa föreskrifna oändlighetspunkter hvilkas konstanter äro på förhand angifna. *Öfversigt af Kungl. Vetenskaps-Akademiens Förhandlingar* 6; 3–16.
1892   Sophie Kovalevsky, Notice Biographique. *AM* 16; 385–390.
1923a   Die ersten 40 Jahre des Lebens von Weierstrass. *AM* 39; 1–57.
1923b   Correspondance de Henri Poincaré et de Felix Klein. *AM* 39; 94–132.
1923c   Weierstrass et Sonja Kowalewsky. *AM* 39; 133–198.
1923d   Briefe von Weierstrass an Paul du Bois–Reymond. *AM* 39; 199–225.
1923e   Briefe von Weierstrass an L. Koenigsberger. *AM* 39; 226–239.
1923f   Briefe von Weierstrass an L. Fuchs. *AM* 39; 246–256.

NEKRASOV, P. A.

1891   On the works of S. V. Kovalevskaya in pure mathematics (Russian). *Matematicheskii Sbornik* 16, No. 1; 31–38.

NEUENSCHWANDER, E.

1981   Studies in the history of complex function theory II: Interactions among the French school, Riemann, and Weierstrass. *Bulletin of the American Mathematical Society* 5, No. 2; 87–105.

NEUMANN, C.

1856   De problemate quodam mechanico, quod ad primam integralium ultraellipticorum classem revocantur. Dissertation (Königsberg).

1871   Nortiz über die elliptischen und hyperelliptischen Integrale. *Mathematische Annalen* III; 611–630.

NIVEN, W. D.

1890   *The Scientific Papers of James Clerk Maxwell.* 2 vols. Cambridge
University Press. Dover reprint, 1965.

NOETHER, E.

1919   Die arithmetische Theorie der algebraischen Funktionen einer Verän-
derlichen, in ihrer Beziehung zu den übrigen Theorien und zu der
Zahlkörpertheorie. *Jahresbericht der deutschen Mathematiker-
Vereinigung* XXVIII; 182–203.

NOVÝ, L.

1971   Paul du Bois-Reymond. *Dictionary of Scientific Biography* IV; 205–206.

OLEINIK, O.

1975   The theorem of S. V. Kovalevskaya and its role in the modern theory
of partial differential equations (Russian). *Matematika v Shkole* No. 5;
5–9.

OSGOOD, W.

1901   Allgemeine Theorie der analytischen Funktionen. *Enzyklopädie der
mathematischen Wissenschaften* II, 2; 5–114.

1928   *Lehrbuch der Funktionentheorie, Fünfte Auflage, Bd. I.* Leipzig: Teubner.

POINCARÉ, H.

1884   Sur la réduction des intégrales abéliennes. *CR* XCIX; 853–855.

1885   Sur l'équilibre d'une masse fluide animée d'un mouvement de rotation.
Bulletin Astronomique 2; 405–413.

1890   Sur le problème des trois corps et les équations de la dynamique. *AM*
13; 1–270.

                Oeuvres. 11 vols. 1951–1956 (volume 3 reprinted from 1934). Paris:
Gauthier-Villar.

POINSOT, L.

1834   *Théorie nouvelle de la rotation des corps*. Présentée à l'Institut, Paris.

POISSON, S. D.

1833   *Traité de mécanique, deuxième édition*. 2 vols. Paris: Bachelier.

RAPPAPORT, K.

1981   Sonya Kovalevsky. *American Mathematical Monthly* 88; 564–574

RICHELOT, F.

1846   Über die Reduction des Integrales $\int [fx \, dx / \sqrt{\pm(1 - x^8)}]$ auf ellip-
            tische Integrale. *JFM* 32; 213–218.

RIEMANN, B.

1851   Grundlagen für eine allgemeine Theorie der Functionen einer verän-
            derlichen complexen Grösse (Inaugural Dissertation, Göttingen) =
            Werke, 3–45.
1854   Über die Hypothesen, welche der Geometrie zu Grunde liegen
            (Habilitationsschrift). *Abhandlungen der Königl. Gesellschaft der
            Wissenschaften zu Göttingen = Werke;* 272–287.
1857   Theorie der abel'schen Functionen. *JFM* 54; 115–155. = *Werke;*
            88–142.
            Über das Potential eines Ringes. *Werke;* 431–436.
            Schwere, Electricität und Magnetismus (K. Hattendorff, ed.) 1876.
            Hannover.
            Gesammelte Mathematische Werke, zweite Auflage (H. Weber, ed.)
            1892. Dover, New York, 1953.

RIESZ, F. and SZ.-NAGY, B.

1955   *Functional Analysis*. Translated from the 2nd French edition by Leo F.
            Boron, New York: Ungar.

ROSENHAIN, G.

1850   Auszug mehrerer Schreiben des Rosenhain an Herrn Professor Jacobi
            über die hyperelliptischen transcendenten. *JFM* 40; 320–360.

RUNGE, C.

1885   Zur Theorie der eindeutigen analytischen Funktionen. *AM* 6; 229–245.

SHTRAIKH, S. YA.

1951   *S. V. Kovalevskaya, Reminiscences and Letters* (Russian). Moscow.

STILLMAN, B.

1978.   *A Russian Childhood*. New York: Springer-Verlag.

STOLETOV, A.

1891   S. V. Kovalevskaya, A biographical essay (Russian). *Mathematicheskii Sbornik* 16; 1–10.

TEE, G.

1977   Sof'ya Vasilievna Kovalevskaya. *Math. Chronicle* 5; 113–139.
1980   Review of *Sofya Kovalevskaya: A Russian Childhood*, by B. Stillman. *Annals of Science* 37; 469–471.

THIMM, W.

1966   Der Weierstrassche Satz der algebraischen Abhängigkeit von abel'schen Funktionen und seine Verallgemeinerungen. In *Behnke & Kopfermann*, 123–153.

THOMAS À KEMPIS, SR. M.

1966   Mathematics and the Nobel Prize. *The Mathematics Teacher* LIX; 667–668.

TISSERAND, F.

1891   *Traité de Mécanique Céleste, t. 2*. Paris: Gauthier-Villars.

VOLTERRA, V.

1892   Sur les vibrations lumineuses dans les milieux biréfringents. *AM* 16; 153–206.

WEBER, H.

1878   Über die Kummersche Fläche vierter Ordnung mit sechzehn Knotenpunkten und ihrer Beziehung zu den Thetafunktionen mit zwei Veränderlichen. *JFM* 84; 332–354.

WEBSTER, A. G.

1894   Letter to the editor. *The Nation* 58, No. 1492; 83.

WEIERSTRASS, K.

1842   Definition analytischer Functionen einer Veränderlichen vermittelst al-
               gebraischer Differentialgleichungen. *Werke* I; 75–84.
1854   Zur Theorie der abel'schen Functionen. *JFM* 47; 289–306 = *Werke* I;
               133–152.
1856a   Über die Theorie der analytischen Facultäten. *JFM* 51;
               1 − 60 = *Werke* I; 153–221 (Small variances between the two texts).
1856b   Theorie der abel'schen Functionen. *JFM* 52; 285–380. Excerpt in
               Werke I; 297–355.
1857   Akademische Antrittsrede. *Monatsberichte der Königlichen Pre-
               ussischen Akademie der Wissenschaften zu Berlin*, 9 July
               1857 = *Werke* I; 223–226.
1861   Über die geodätischen Linien auf dem dreiaxigen Ellipsoid. *Mon-
               atsberichte der Königlichen Preussischen Akademie der Wissenschaften
               zu Berlin* (1861) 986–997 = *Werke* I; 257–266.
1869   Über die allgemeinsten eindeutigen und 2n-fach periodischen Func-
               tionen von n Veränderlichen. *Monatsberichte der Königlichen Pre-
               ussischen Akademie der Wissenschaften zu Berlin* (1869)
               853–857 = *Werke* II; 45–48.
1870   Über das sogenannte Dirichletsche Princip. Paper read at a meeting of
               the Königliche Preussische Akademie der Wissenschaften zu Berlin, 14
               July 1870 = *Werke* II; 49–54.
1880a   Über einen functionentheoretischen Satz des Herrn G. Mittag-Leffler.
               *Monatsberichte der Königlichen Preussischen Akademie der Wis-
               senschaften zu Berlin* (1880) 707–717 = *1886;* 53–66 = *Werke* II;
               189–199.
1880b   Untersuchungen über die 2*r*-fach periodischen Functionen von *r* Verän-
               derlichen. *JFM* 89; 1–8 = *Werke* II; 124–133.
1886   *Abhandlungen aus der Functionenlehre.* Berlin:Springer.

               Vorlesungen über die Theorie der abelschen Transcendenten = *Werke*
               IV.

               Vorlesungen über Anwendungen der elliptischen Functionen = *Werke*
               VI.

               Zur Dioptrik. Presented at the 22 September 1856 meeting of the math-
               ematical section of the Union of German Natural Scientists and Physi-
               cians in Vienna. *Werke* III; 175–178.

               Mathematische Werke. 7 vols. 1894–1927. Hildesheim: Georg Olms
               Verlagsbuchhandlung. Johnson Reprint Corporation, New York.

WEIL, A.

1982   Mittag-Leffler as I remember him. *AM* 148; 9–13.

YOUNG, G. C.

1916   On the derivates of a function. *Proceedings of the London Mathematical Society* (2) 15; 360–384.

ZHUKOVSKY, N.

1891   On the works of S. V. Kovalevskaya in applied mathematics (Russian). *Matematicheskii Sbornik* 16; 11–30.

1897   Geometrische Interpretation des von Sophie Kowalevski behandelten Falles der Bewegung eines schweren starren Körpers um einen festen Punkt. *Jahresbericht der deutschen Mathematiker-Vereinung* IV: 144–150.

ZIGELAAR, A.

1980   How did the wave theory of light take shape in the mind of Christiaan Huyghens? *Annals of Science* 37; 179–187.

Index

Previously published by Springer-Verlag

A Russian Childhood
by Sofya Kovalevskaya
Introduced, translated, and edited by Beatrice Stillman

"Not many mathematicians are known for literary talents—Kovalevskaya was
not only a good enough mathematician to win the Prix Bordin, but an able
writer, as demonstrated by this enjoyable book."—Mathematics Magazine

". . . an engaging introduction to her life and work, ... The memoir in
particular evinces her deep modesty, integrity, and psychological insight."—
Choice

"It was not an easy task to render into literary English the extraordinarily
beautiful prose of the original. Here Stillman has been successful."—Canadian
Slavonic Papers

"Stillman's translation is fluent and very accurate, ..."—Annals of Science

Kovalevskaya's childhood reminiscences are a delightful piece of cultural and
social history, capturing the period of the rise of radical political groups and the
emancipation of serfs of the late 1800s. Ranging from minute descriptions of
earliest memories to her friendship with Dostoevsky during her teenage years,
the book is a thoroughly absorbing panorama that remains as alive today as it
was in 1889 when it was first published.

1978.   272 pp.   6 illus.   ISBN 0-387-90348-8

 Springer-Verlag New York Berlin Heidelberg Tokyo